Implantable Microdevices

Implantable Microdevices

Special Issue Editors

Wen Li
Zhen Qiu

MDPI • Basel • Beijing • Wuhan • Barcelona • Belgrade

Special Issue Editors

Wen Li
Michigan State University
USA

Zhen Qiu
Michigan State University
USA

Editorial Office
MDPI
St. Alban-Anlage 66
4052 Basel, Switzerland

This is a reprint of articles from the Special Issue published online in the open access journal *Micromachines* (ISSN 2072-666X) from 2018 to 2019 (available at: https://www.mdpi.com/journal/micromachines/special_issues/Implantable_Microdevices)

For citation purposes, cite each article independently as indicated on the article page online and as indicated below:

LastName, A.A.; LastName, B.B.; LastName, C.C. Article Title. *Journal Name* **Year**, *Article Number*, Page Range.

ISBN 978-3-03921-660-4 (Pbk)
ISBN 978-3-03921-661-1 (PDF)

Contents

About the Special Issue Editors . . . . **vii**

Wen Li and Zhen Qiu
Editorial for the Special Issue on Implantable Microdevices
Reprinted from: *Micromachines* **2019**, *10*, 603, doi:10.3390/mi10090603 . . . . **1**

Qi Zeng, Saisai Zhao, Hangao Yang, Yi Zhang and Tianzhun Wu
Micro/Nano Technologies for High-Density Retinal Implant
Reprinted from: *Micromachines* **2019**, *10*, 419, doi:10.3390/mi10060419 . . . . **3**

Liguo Qin, Mahshid Hafezi, Hao Yang, Guangneng Dong and Yali Zhang
Constructing a Dual-Function Surface by Microcasting and Nanospraying for Efficient Drag
Reduction and Potential Antifouling Capabilities
Reprinted from: *Micromachines* **2019**, *10*, 490, doi:10.3390/mi10070490 . . . . **25**

Ye Xi, Bowen Ji, Zhejun Guo, Wen Li and Jingquan Liu
Fabrication and Characterization of Micro-Nano Electrodes for Implantable BCI
Reprinted from: *Micromachines* **2019**, *10*, 242, doi:10.3390/mi10040242 . . . . **40**

Amirhossein Hajiaghajani and Seungyoung Ahn
Single-Sided Near-Field Wireless Power Transfer by A Three-Dimensional Coil Array
Reprinted from: *Micromachines* **2019**, *10*, 200, doi:10.3390/mi10030200 . . . . **49**

Yi Fan, Xiongying Liu, Jiming Li and Tianhai Chang
A Miniaturized Circularly-Polarized Antenna for In-Body Wireless Communications
Reprinted from: *Micromachines* **2019**, *10*, 70, doi:10.3390/mi10010070 . . . . **58**

Rongqiang Li, Bo Li, Guohong Du, Xiaofeng Sun and Haoran Sun
A Compact Broadband Antenna with Dual-Resonance for Implantable Devices
Reprinted from: *Micromachines* **2019**, *10*, 59, doi:10.3390/mi10010059 . . . . **69**

Houguang Liu, Hehe Wang, Zhushi Rao, Jianhua Yang and Shanguo Yang
Numerical Study and Optimization of a Novel Piezoelectric Transducer for a Round-Window
Stimulating Type Middle-Ear Implant
Reprinted from: *Micromachines* **2019**, *10*, 40, doi:10.3390/mi10010040 . . . . **76**

Jae-Ho Lee and Dong-Wook Seo
Development of ECG Monitoring System and Implantable Device with Wireless Charging
Reprinted from: *Micromachines* **2019**, *10*, 38, doi:10.3390/mi10010038 . . . . **92**

**Yun-Ting Chen, Kun-Ying Yeh, Szu-Han Chen, Chuang-Yin Wang, Chao-Chi Yeh,
Ming-Xin Xu, Shey-Shi Lu, Yunn-Jy Chen and Yao-Joe Yang**
Tongue Pressure Sensing Array Integrated with a System-on-Chip Embedded in a Mandibular
Advancement Splint
Reprinted from: *Micromachines* **2018**, *9*, 352, doi:10.3390/mi9070352 . . . . **107**

About the Special Issue Editors

Wen Li received the B.S. degree in Material Science and Engineering from Tsinghua University, Beijing, and the M.S. and Ph.D. degrees both in Electrical Engineering from California Institute of Technology, Pasadena, in 2004 and 2009, respectively. Currently she is an associate professor in the Departments of Electrical and Computer Engineering at Michigan State University, East Lansing. Her research interests include bioMEMS, neuroprosthetic devices, micro/nanosensors, nanoelectronics, polymer microfabrication process development, and microsystem integration and packaging technologies. Li received an NSF CAREER Award (2011), the Best Application Paper Award at 3M-NANO (2011), and the Best Paper Award at International Neurotechnology Consortium (2013). She serves as an associate editor of Transaction of Biomedical Engineering and Micro & Nano Letters, and a technical program committee member of IEEE EMBS Wearable Biomedical Sensors and Systems Technical Committee. She is an IEEE senior member, and a member of the American Chemical Society and Biomedical Engineering Society.

Zhen Qiu received the Bachelor Degree in the Department of Precision Instruments, Tsinghua University, Beijing, China, and the Ph.D. degree from the Department of Biomedical Engineering, University of Michigan-Ann Arbor, MI, USA, in 2004, 2013, respectively. He finished his post-doctoral training in the Department of Radiology and Molecular Imaging Program Stanford (MIPS), School of Medicine, Stanford University, CA. He is currently an Assistant Professor in the Department of Biomedical Engineering, Institute for Quantitative Health Science and Engineering, Michigan State University, MI. His research interests include biomedical optics, MEMS/MOEMS, multi-modal targeted imaging using novel contrast agents, wearable and implantable medical devices, ultrafast laser applications. Professor Qiu is currently holding the member of SPIE and OSA.

 *micromachines*

Editorial

Editorial for the Special Issue on Implantable Microdevices

Wen Li [1,2,3,*] **and Zhen Qiu** [1,2,3,*]

1 Department of Electrical and Computer Engineering, Michigan State University, East Lansing, MI 48824, USA
2 Department of Biomedical Engineering, Michigan State University, East Lansing, MI 48824, USA
3 Institute for Quantitative Health Science and Engineering, Michigan State University, East Lansing, MI 48824, USA
* Correspondence: wenli@msu.edu (W.L.); qiuzhen@egr.msu.edu (Z.Q.)

Received: 10 September 2019; Accepted: 11 September 2019; Published: 12 September 2019

Implantable microdevices, providing accurate measurement of target analytes in animals and humans, have always been important in biological science, medical diagnostics, clinical therapy, and personal healthcare. Recently, there have been increasing unmet needs for developing high-performance implants that are small, minimally-invasive, biocompatible, long-term stable, and cost-effective. Therefore, this special issue aims to bring together state-of-the-art research and development contributions that address key challenges and topics related to implantable microdevices.

There are eight research articles and one review article published in this special issue, covering a wide range of topics: implantable sensors for detecting biophysiological and biophysical signals [1,2]; novel transducers for stimulation [3]; miniaturized antennas and biotelemetries for wireless implants [4–6] medical electronics and system-on-chip (SoC) [1,7]; and nano/microfabrication technologies for implantable microdevices [2,8,9]. On the device side, Chen et al. reported a tongue pressure sensing platform that integrated interdigitated pressure sensing electrodes with SoC on flexible polymer films. The sensor achieved high sensitivities at the low-pressure region of below 10 kPa and good stability in air and in deionized water. Xi et al. fabricated an ultramicron silver needle electrode (7.9 μm in diameter) using a cost-effective, laser-assisted pulling method [2]. Also published is a new piezoelectric transducer implant proposed by Liu et al. for stimulation of the round window membrane of the middle ear with significantly reduced power consumption and improved output quality [3]. On the system side, Lee et al. presented a highly efficient, wireless ECG monitoring system that integrated ECG sensor, temperature sensor, microcontroller unit, radio-frequency transceiver, and wireless power transmission unit on a single printed circuit board for continuous monitoring of cardiac arrhythmias [7].

Inductive telemetries have been widely used in biomedical implants for wireless power delivery and data communication. This special issue includes multiple designs of wireless antennas that have been explored by different groups. In particular, Hajiaghajani et al. proposed a three-dimensional coil array based on a generalized Halbach array for wireless powering of biomedical instruments within a large area of 144 cm^2 while eliminating the need for traditional magnetic shielding [4]. In another design, Fan et al. designed a miniaturized circularly polarized antenna for in-body wireless communication with broadened bandwidth [5]. Li et al. also proposed a compact broadband implantable antenna with dual resonance bandwidth of 52 MHz at low return loss of −10 dB. Packaging and surface fouling are also long-standing challenges of biomedical implants. Zeng et al. presented a comprehensive review article that summarizes the micro/nanotechnologies and material considerations for high-density implantable packaging and flexible microelectrode arrays used in artificial retinal prostheses [8]. To reduce the surface fouling effect, Qin et al. proposed a dual bio-inspired shark-skin and lotus-structure surface with a unique combination of superhydrophobic, antidrag, and self-cleaning properties [9].

We would like to take this opportunity to thank all the authors for submitting their papers to this special issue. We also want to thank all the reviewers for dedicating their time and helping to improve the quality of the submitted papers.

Conflicts of Interest: The author declares no conflict of interest.

References

1. Chen, Y.T.; Yeh, K.Y.; Chen, S.H.; Wang, C.Y.; Yeh, C.C.; Xu, M.X.; Lu, S.S.; Chen, Y.J.; Yang, Y.J. Tongue Pressure Sensing Array Integrated with a System-on-Chip Embedded in a Mandibular Advancement Splint. *Micromachines* **2018**, *9*, 352. [CrossRef] [PubMed]
2. Xi, Y.; Ji, B.; Guo, Z.; Li, W.; Liu, J. Fabrication and Characterization of Micro-Nano Electrodes for Implantable BCI. *Micromachines* **2019**, *10*, 242. [CrossRef] [PubMed]
3. Liu, H.; Wang, H.; Rao, Z.; Yang, J.; Yang, S. Numerical Study and Optimization of a Novel Piezoelectric Transducer for a Round-Window Stimulating Type Middle-Ear Implant. *Micromachines* **2019**, *10*, 40. [CrossRef] [PubMed]
4. Hajiaghajani, A.; Ahn, S. Single-Sided Near-Field Wireless Power Transfer by A Three-Dimensional Coil Array. *Micromachines* **2019**, *10*, 200. [CrossRef] [PubMed]
5. Fan, Y.; Liu, X.; Li, J.; Chang, T. A Miniaturized Circularly-Polarized Antenna for In-Body Wireless Communications. *Micromachines* **2019**, *10*, 70. [CrossRef] [PubMed]
6. Li, R.; Li, B.; Du, G.; Sun, X.; Sun, H. A Compact Broadband Antenna with Dual-Resonance for Implantable Devices. *Micromachines* **2019**, *10*, 59. [CrossRef] [PubMed]
7. Lee, J.H.; Seo, D.W. Development of ECG Monitoring System and Implantable Device with Wireless Charging. *Micromachines* **2019**, *10*, 38. [CrossRef] [PubMed]
8. Zeng, Q.; Zhao, S.; Yang, H.; Zhang, Y.; Wu, T. Micro/Nano Technologies for High-Density Retinal Implant. *Micromachines* **2019**, *10*, 419. [CrossRef] [PubMed]
9. Qin, L.; Hafezi, M.; Yang, H.; Dong, G.; Zhang, Y. Constructing a Dual-Function Surface by Microcasting and Nanospraying for Efficient Drag Reduction and Potential Antifouling Capabilities. *Micromachines* **2019**, *10*, 490. [CrossRef] [PubMed]

Review

Micro/Nano Technologies for High-Density Retinal Implant

Qi Zeng [1,†], **Saisai Zhao** [1,†], **Hangao Yang** [1], **Yi Zhang** [2] **and Tianzhun Wu** [1,*]

1   Shenzhen Institutes of Advanced Technology (SIAT), Chinese Academy of Sciences (CAS), Shenzhen 518055, China; qi.zeng@siat.ac.cn (Q.Z.); ss.zhao@siat.ac.cn (S.Z.); hg.yang@siat.ac.cn (H.Y.)
2   Shenzhen CAS-Envision Medical Technology Co. Ltd., Shenzhen 518100, China; zhangyi@cas-envision.com
*   Correspondence: tz.wu@siat.ac.cn; Tel.: +86-755-8639-2339
†   These authors contributed equally to this work.

Received: 22 May 2019; Accepted: 21 June 2019; Published: 22 June 2019

Abstract: During the past decades, there have been leaps in the development of micro/nano retinal implant technologies, which is one of the emerging applications in neural interfaces to restore vision. However, higher feedthroughs within a limited space are needed for more complex electronic systems and precise neural modulations. Active implantable medical electronics are required to have good electrical and mechanical properties, such as being small, light, and biocompatible, and with low power consumption and minimal immunological reactions during long-term implantation. For this purpose, high-density implantable packaging and flexible microelectrode arrays (fMEAs) as well as high-performance coating materials for retinal stimulation are crucial to achieve high resolution. In this review, we mainly focus on the considerations of the high-feedthrough encapsulation of implantable biomedical components to prolong working life, and fMEAs for different implant sites to deliver electrical stimulation to targeted retinal neuron cells. In addition, the functional electrode materials to achieve superior stimulation efficiency are also reviewed. The existing challenge and future research directions of micro/nano technologies for retinal implant are briefly discussed at the end of the review.

Keywords: retinal implant; high-density; implantable packaging; microelectrode array; coating

1. Introduction

Implanted microelectronic device has enormous applications in biomedical and animal studies, and one of the emerging applications is the neural interface micro-devices for neural activity monitoring and modulation [1–3]. Since the first pacemaker implantation in 1958, many implantable medical devices such as deep brain stimulators, cochlear implants, and blood pressure sensors have been developed and commercialized [4,5]. These implantable medical devices typically have only 4–26 feedthroughs (or independently controlled channels), which connect the electrode array and receiver coil to the stimulator. However, higher feedthroughs within a limited space are needed for supporting more complex electronic systems and precise neural modulations. One good example is retinal prostheses to restore the basic visual functions for patients suffering from severe retinal degeneration disease including retinal pigmentosa (RP) and dry age-related macular degeneration (AMD), also known as the bionic eye, retinal prosthesis, or visual prosthesis [6,7]. AMD occurs predominately in the aged, affecting 30–50 million people globally and more than 4 million in China alone, while RP afflicts mainly children and young adults, affecting 1.5 million people globally [7–10].

With rapid advancements in technology and engineering for retinal implant, the first commercial retinal prostheses, the Argus® II retinal prosthesis system (referred to as simply Argus II) with 60 channels from Second Sight Medical Products Inc., has obtained the European Union approval (CE Mark) in 2011, and the U.S. Food and Drug Administration (FDA) with the humanitarian device exemption in 2013, has been commercially implanted in hundreds of patients [8,9]. The working principle and clinical results of Argus II have been detailed in many review papers [7,10–12], and its illustration is shown in Figure 1a,b. In 2013, Alpha IMS from Retina Implant AG became the second CE certified wireless retina implant in Europe, featured as a subretinal device implanted under the damaged retina to take advantage of the signal processing capabilities of intact bipolar cells in the retina [13]. The implant uses a 3-mm square microphotodiode array (MPDA) chip consisting of 1500 photodiodes to amplify pixels instead of the eye's photoreceptors. Another method with the retinal implantation site between the sclera and choroid, namely suprachoroid implant, was also adopted with relatively simple and less invasive surgery [14,15]. Figure 1c shows the major approaches to retinal implants depending on the implantation sites. The intraretinal neurons largely retain the ability to transmit signals despite reorganization and cell loss. The images captured from the visual field through the image acquisition device are translated into stimulation patterns for the electrode array placed near the retina. An electrochemical interface can be formed with physiological saline. The current delivered into the extracellular region causes charge redistribution on the membrane of retinal neurons, and an action potential is triggered when the membrane depolarization exceeds a threshold. For example, a negative charge accumulates outside the membrane under the electrode during cathodic stimulation, driving the intracellular positive charge moving to the electrode/neuron interface from the adjacent compartments, and leading to a strong membrane depolarization closer to the electrode and a weaker hyperpolarization far away. Anodic stimulation is the opposite [7,16]. Besides RP patients, more than 600 AMD patients received an entirely different type of implant to restore their poor vision: the implantable micro-telescope (IMT) from Vision Care of Saratoga, California [17]. The pea-sized IMT is implanted behind the iris of eye. The loss of central vision can be magnified 2.7 times by the telescope and projected onto the healthy part of the retina surrounding the scarring.

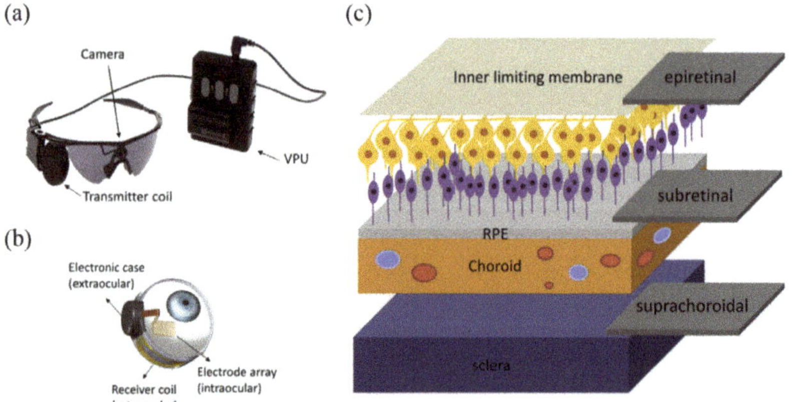

Figure 1. (**a**) External and (**b**) implant part of the Argus II system [7]; (**c**) illustration of the implantation sites of the visual cortex, epiretinal, subretinal, and supra-choroidal prostheses [7]. Reproduced with permission from [7], published by Elsevier, 2016.

There is a long-lasting effort to increase the resolution and visual acuity of retinal implant [6,18,19]. Previous simulation has shown that a super-high density of 600–1000 stimulus electrodes (pixels) is required to help blind people with face recognition and reading large text in newspapers [20]. For this purpose, high-density implantable packaging and flexible micro electrode array (fMEA) as well as high-performance coating materials for stimulation are crucial to achieve high resolution, which will be the focus in this review.

Considering the security issues of multilayer in chronic use, all implants must be stable, biocompatible, and ensure minimal immunological reactions [21–23]. Most active implant devices use silicon (Si)-based electronic devices to perform the functions of sensing or treatment. Due to the biological incompatibility of Si substrate, higher density input/output (I/O) pins and peripheral circuit communication are required. There is a great demand in shrinking the implant size while increasing feedthrough counts [24–27], which has brought great challenges for packaging. Conventional direct packaging using plastic, metal, or ceramic cannot guarantee the long-term moisture-free environment for high-channel-count electronic devices. When water vapor inside the packaging exceeds a threshold value, dewing may occur, and the water droplets together with residual ions will form an electrolytic environment, leading to corrosion, delamination, cracking, or other degradations of the internal components. Therefore, electronic packaging with high density, air tightness, and biocompatibility is an inevitable choice, to build a seal cavity with good mechanical performance and biocompatibility [28,29].

For precise neuronal stimulation, smaller and more sophisticated microelectrode arrays are of increasing interests [30,31]. Considering the major target cells for stimulation, retinal ganglion cell (RGC) has the soma size in 5–20 μm range, so the ideal electrode size should be similar to this size to facilitate one-on-one stimulation. Actually, the optimal size of electrode suggested by the electrophysiologic experiments should be 10 to 20 μm, and the spacing between 20 and 60 μm [32], which are far less than those in current commercial products (for example, the electrodes of the Argus II are 200 μm in diameter and 300 μm in spacing). The decreasing size of electrodes will increase the impedance and improve the signal-to-noise ratio (SNR) dramatically. Moreover, the fabrication process of high-density fMEA in limited size is challenging because of low yield and easy delamination. Therefore, it is critical to study the electrode fabrication process, materials, and surface modification.

For those tiny microelectrodes, high capacitance and charge injection capability (CIC), low electrical impedance, as well as the biocompatibility, are of great importance to reduce power consumption, lower heat dissipation, and ensure safe stimulation [33,34]. Many researchers have attempted to increase the effective surface area of microelectrodes of the given size to improve the charge delivery characteristics [35,36]. A fundamental consideration for the retinal prosthesis is the selection of proper electrode materials which should exhibit biocompatibility and be able to deliver sufficient charge within safe limits. The requirement of porous and rough coatings on the bare microelectrodes has brought opportunities for the application of various nanomaterials in the retinal implant.

It is worth to note high-density stimulation is not equal to high visual acuity (resolution), which requires a lot of factors in both hardware (electrodes and coatings) and software (stimulation strategies based on surgical results). For example, Argus II with 60 channels could only restore the patient's visual acuity to 200/1262 without amplification [37], and Alpha-IMS with 1500 photoelectric pixels had, at its best report, up to 20/546 [38], only slightly higher than Argus II and far below the expected visual acuity. Here we will summarize only the recent progress in the device's fabrication towards high-resolution retinal implant. The survey of stimulation strategies for retinal prosthesis can be found in other reviews [7,39].

2. Packaging and Integration

The packaging of implantable biomedical components has been one of the greatest challenges for chronic implantation. All the materials exposed must be highly biocompatible, and also highly inert to the erosion of body fluids. Now there are two typical kinds of packaging approaches: hard packaging and soft packaging. The former has a hard shell or capsule made of metal, glass, or ceramics, which allows multiple connections to the components of the basement or integrated into the system of internal packages. It has been widely used in commercial products and allows mass production [40], but the manufacturing costs and risk of failure are rapidly increasing for high-density feedthroughs. Many efforts have been dedicated to increase package feedthrough and reliable bonding/joint using new processes and materials [26–29]. The second option is using one or several layers of biocompatible soft films as a hermetical coating, which has the advantages of being a small-size, light-weight, low-cost, and simple process with high flexibility [41], and has attracted a lot of research interest as an emerging technology enabling high-density, ultra-small medical implants.

2.1. Hard Packaging and Integration

A typical hard packaging for retinal implant is the ceramic/metal composite can for Argus II. The implantable packaging includes three steps: (a) the fabrication of ceramic substrate with Pt feedthroughs, (b) achieving brazed joint with a titanium (Ti) ring, and (c) laser welding with a Ti cap. The illustration of a patented method for the fabrication process is shown in Figure 2a [42]. Multiple blind holes inside a green, or unfired, ceramic sheet were formed firstly, solid wires (or pins) like platinum (Pt) were inserted, followed by sintering treatment for the shrinkage of ceramic and removing the extra materials, so that the lower ends of the wires were flush with the lower surface of the finished sheet.

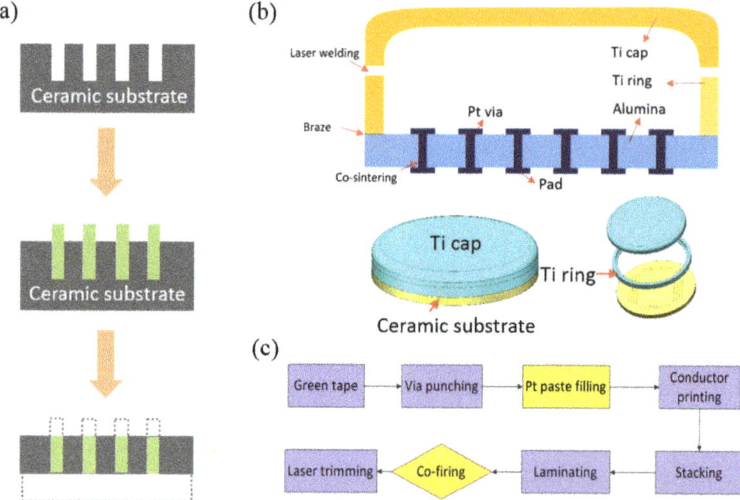

Figure 2. (**a**) Illustration of the package fabrication process described by a patent from Second Sight; (**b**) the cross-sectional (top) and 3D illustration (bottom) of implantable body for retinal implant with 100+ channel, and (**c**) the major process flow for platinum/alumina composite substrate in our group.

Our group in Shenzhen Institutes of Advanced Technology (SIAT), Chinese Academy of Sciences (CAS) also adapted the hard packaging approach to host the retinal implant with more than 130 feedthroughs, but the microfabrication process was optimized in a different way. The packaging mainly consists of alumina/platinum (Al_2O_3/Pt) substrate, a Ti ring, and a Ti cap, which allows 100–500 feedthroughs in 1 cm^2 and shows the best cytotoxicity grade (Grade 0) [43]. As shown in Figure 2b, Pt vias were embedded in Al_2O_3 sheets and covered with Pt pads for electric connection. The Ti ring was brazed with Al_2O_3 and bonded to Ti cap using laser welding. The whole packaging body housed circuits or power inside to be isolated from gas and water penetration.

As shown in Figure 2c, high-purity Al_2O_3 power and its organic adhesives were mixed into emulsion and laminated into a thin green tape. Then mechanical milling was conducted to make micrometer holes in the green tape. Pt paste was filled into the holes typically using thick-film printing technology. After that, Pt or gold (Au) line trace and pad can be patterned on the tap using screen printing technology with the aid of a stainless-steel mesh. Several patterned green tapes were aligned and stacked with uniform pressure using the commercially available isopressing equipment, followed by co-sintering of metal paste and Al_2O_3. Finally, the substrate was diced into desired pieces by laser cutting. The Ti ring was then brazed with Al_2O_3 and inserted between the Ti ring and the ceramic substrate with the aid of a thin braze alloy sheet. After mounting the integrated circuit (IC) and discrete components using surface mounted technology (SMT), the Ti ring was sealed with Ti cap using laser welding. Picosecond laser is used since it allows metal melting immediately at low-temperatures

(<100 °C) to fulfill the metal interface, without additional intermediate layer. All the joint parts have reached an ultra-low level of 10^{-10} Pa·m^3/s for helium (He), which is at least 10 times lower than the 10-year implant requirement of FDA standard (10^{-9} Pa·m^3/s), and they also exhibit excellent biocompatibility (Grade 0) using L929 cell line.

Another report on high-density ceramic/Pt ceramic packaging with integrated electrodes is from the University of New South Wales [44,45], up to 2500 channels per cm^2 (Figure 3a). It is featured with chip-scale packages with integrated electronics, and the outer layer of microelectrodes can directly interface with neural tissues. The charge injection limit of these electrodes and stability under stimulation were explored, showing excellent stability under stimulation with more than 1.8 billion pulses and the electrodes have improved charge transfer properties when compared to machined Pt microelectrodes.

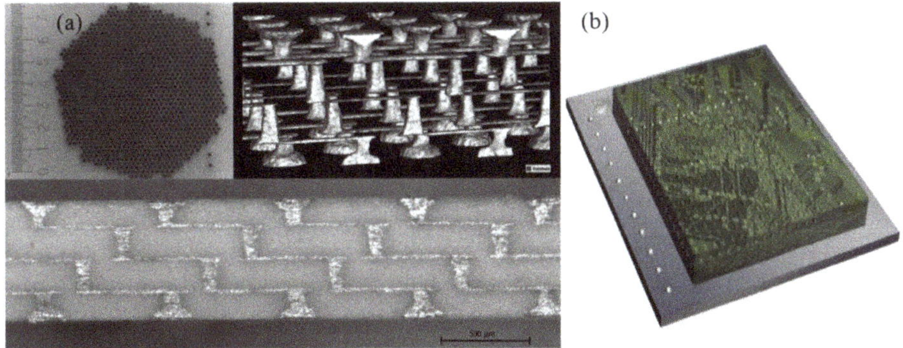

Figure 3. (**a**) The high-density electrode array and corresponding cross-section of hermetic feedthroughs produced with four layers, Reproduced with permission from [44], published by Wiley Online Library, 2014; (**b**) illustration of a high-density array of diamond feedthrough and electrode with 256 channels, Reproduced with permission from [25], published by Elsevier, 2014.

There are some other materials available for the substrate, including diamond and glass. Figure 3b shows an illustration of a high-density array of diamond feedthroughs and electrodes reported by the Bionic Vision Australia (BVA) group in the University of Melbourne [25]. The 256-channel feedthrough array was constructed from a kind of polycrystalline diamond with electrically insulating diamond substrate, containing many electrically conducting nitrogen doped ultra nano-crystalline diamond (N-UNCD) feedthroughs. N-UNCD has appropriate electrochemical characteristics to act as a neural stimulation material [26,27]. Although the growth and microfabrication of diamond materials was quite complicated and expensive, this method has several advantages. First, diamond can minimize the possibility of feedthrough failure by materials mismatch, which increases the reliability. Also, diamonds exhibit excellent stability, good biocompatibility [28,42,46], and superb biochemical stability [25], offering the prospect of a long-lasting implant.

Recently, through glass via (TGV) technology was reported from Waseda University as shown in Figure 4. Similar to through silicon via (TSV) technology, TGV has several advantages, including high wiring density, lower energy consumption, and fast signal speed, compared with conventional wire bonding methods [46]. Air leaks from the TGV area are minimized by sandwiching the TGV between two Au bumps fabricated simultaneously with the TGV by a simple filling process, therefore the throughput is high compared with conventional through via technologies.

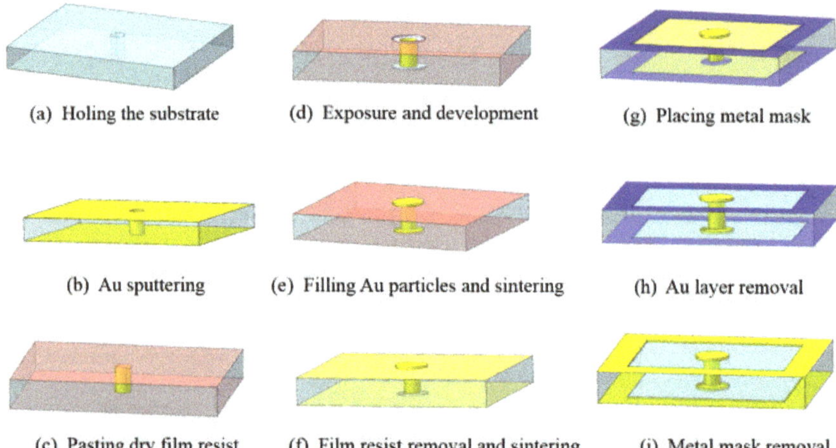

(a) Holing the substrate

(d) Exposure and development

(g) Placing metal mask

(b) Au sputtering

(e) Filling Au particles and sintering

(h) Au layer removal

(c) Pasting dry film resist

(f) Film resist removal and sintering

(i) Metal mask removal

Figure 4. Fabrication process of the base wafer: (**a**) Holing the substrate, (**b**) Au sputtering, (**c**) pasting dry film resist, (**d**) exposure and development, (**e**) filling Au particles and sintering, (**f**) film resist removal and sintering, (**g**) placing metal mask, (**h**) Au layer removal, and (**i**) metal mask removal. Reproduced with permission from [46], published by IOPScience, 2016.

2.2. Soft Packaging and Integration

Soft packaging for medical implants borrowed the idea from the film packaging used for light emitting diode (LED) and liquid crystal display (LCD). At present, the most commonly used film packaging is the combination of multi-layer film to form a packaging barrier layer, also known as Barix packaging technology [47], requiring simple preparation process and low cost. Among the available polymer materials [41], parylene-C is the most commonly used due to its wide applications and comprehensive advantages. Parylene-silicone-parylene film has been demonstrated for 256-channel retina implants [48] and is expected to work equivalently for 7 years in an accelerated test [49]. Our group used ultra-thin (3.1 μm) and biocompatible organic/inorganic composite film based on room-temperature and conformal deposition of Al_2O_3 and parylene C, as shown in Figure 5a [50]. The water penetration route of such film will be significantly extended by steering the penetration path to the organic/inorganic interface, so that combining only 5 layers can significantly improve the lifetime for implantation. Its leakage rate was lower than 1×10^{-10} Pa·m^3/s, with 58 days of the active soaking test for a commercial humidity sensor with this composite film at 87 °C (Figure 5b,c), expected to have the life time longer than 5 years [51].

In addition, liquid crystal polymer (LCP) is a new emerging material due to its high strength, high modulus, ultra-low water absorption, and low expansion coefficient [52–54]. The completed LCP-based retinal prosthetic device is shown in Figure 5d. The device has a circular package accommodating the electronics with a 14 mm diameter and 1.3 mm maximum thickness for its crescent-shaped cross section, which can be conformally attached on the eyeball. The electrode part to be inserted into the retina has a thickness of 30 μm after the laser-thinning process which etched away the LCP starting from a 350 μm thickness and is precured to fit the eye-curvature. This LCP-based retinal prosthesis weighs only 0.38 g, less than a tenth of conventional implantable devices with a metal package. Considering that the weight of an eyeball is about 5 g, this weight reduction is a significant improvement in patients' discomfort as well as implantation stability.

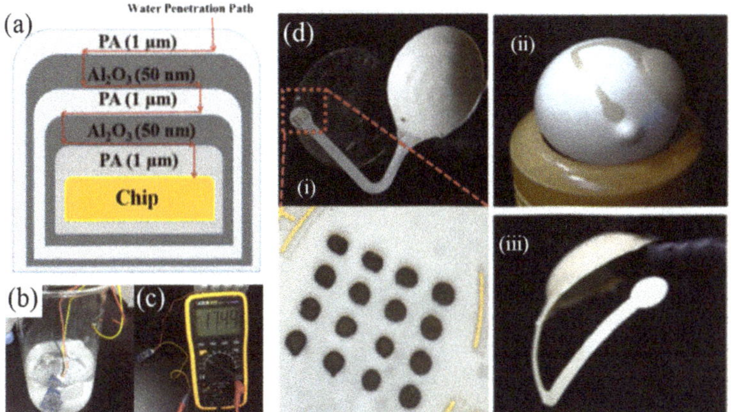

Figure 5. (**a**) Schematic diagram of the five-layer PA/Al$_2$O$_3$/PA/ Al$_2$O$_3$/PA film on a sensor IC; (**b**) active soaking test for the film-coated humidity sensor at 87 °C; (**c**) measurement setting of the humidity sensor after active soaking test; (**d**) fabricated LCP-based retinal prosthesis: (i) comparison with a dime and the inner surface, and magnification of the retinal electrode array coated by iridium oxide, (ii) the device on a model eye showing conformal attachment, (iii) electrode part was precurved to fit the eye-curvature. Reproduced with permission from [55], published by IEEE, 2015.

3. Microelectrode Array for Retinal Implant

In retinal prosthesis, fMEAs are the critical interfaces between the implant system and the tissue that deliver charge-balanced electrical stimulation to targeted retinal neuron cells [56]. Depending on the implant sites, there three main three types of retinal electrode array as shown in Figure 1c, i.e., an epiretinal one on the top surface of the retina, a subretinal one between the retina and retinal pigment epithelium (RPE)/choroid, and a supra-choroidal one on the posterior scleral surface. The following sections for the three types are reviewed in detail.

3.1. Epiretinal Electrodes

Early experiments demonstrated that electrical stimulation could restore visual signals. A single needle was placed onto the retinal surface to simulate the retina by Humayun and co-workers, shown in Figure 6a [57]. The pixels increased for Argus I (16 electrodes) and Argus II (60 electrodes) (Figure 6b,c). The Argus II had been implanted in 30 patients (1 with choroideremia and 29 with RP) in the United States and Europe between 2007 and 2009, all of them could perceive light when given electrical stimulation. The implant is very expensive: about $150,000 without the surgery fee [7,58]. Another typical epiretinal implant is the 49-electrode epiretinal device from Intelligent Medical Implants (IMI) as shown in Figure 6d. Unlike Argus I and II, the power for the IMI retinal stimulator is provided through a radio frequency (RF) links, while the stimulation data could be transmitted via an infrared (IR) optical link to decouple the energy and data interference. It has been demonstrated that the implant did not cause tissue damage in the eye after implanted in 4 RP subjects for over 9 months [59]. EPIRET3 is the latest retinal model from Epi-Ret that the fits entirely within the eyeball, which eliminates the need to suture any component. It contains 25 protruding electrodes (100 μm in diameter), arranged in a hexagonal array, resulting in lower thresholds by improving contact with the retina. It was implanted in six RP patients in 2006 and all subjects reported visual sensations after implantation for 4 weeks [60]. In 2016, Pixium Vision in France commercialized a 150-electrode product (IRIS II) and obtained the CE mark [61,62]. Our group also developed the 126-electrode prototype based on polyimide (PI) substrate and Ti/Pt metal layers [63] and inserted the in the eyes of mini-pigs for 3 months with good biocompatibility (Figure 6e). We also reported fMEA of 1025 electrodes with improved adhesion and reduced impedance (Figure 6f) [64,65]. The new-generation retina implant developed by our group will be more affordable in the future.

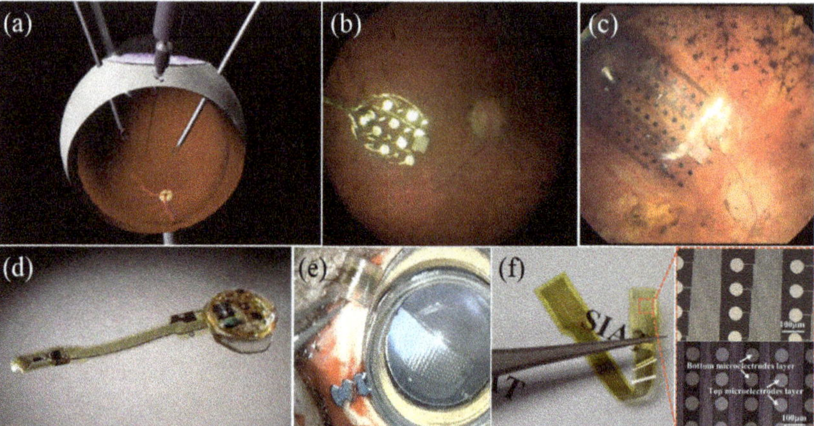

Figure 6. (**a**) The configuration of one of the very first patient tests [57]; (**b**) implanted 16-channel electrode array of Argus I [7]; (**c**) electrode array of Argus II implant containing 60 electrodes [66]; (**d**) an epiretinal stimulator with a thin-film polyimide cable of gold traces [59]; (**e**) 126-channel electrode implanted in the eyes of mini-pigs by our group; (**f**) 1025-channel electrode fabricated by our group [64,65]. Reproduced from the mentioned references with permission from the related journals.

3.2. Subretinal Electrodes

Compared to epiretinal implantation, subretinal implantation enables much higher stimulation density, however, the fMEA is highly integrated with photovoltaic circuits driven only by light inside the eyeball and has the risk of heat dissipation. There still exists disputation as to whether it could result in the atrophy around the implant area in the long term [7]. Artificial silicon retina (ASR) from Optobionics (a company from Chicago) was a typical implant of this kind with a 2-mm diameter silicon-based device, containing about 5000 microelectrodes tipped microphotodiode array (MPDA). 10 RP patients were implanted with ASR and 6 of them had vision during the 7-year follow-up observation [67]. Alpha IMS, developed by Retina Implant AG in Germany, not only used a MPDA for wireless light detection and current generation, but also included wired circuits to amplify the photocurrents to overcome the difficulty of low photovoltaic efficiency [10]. As shown in Figure 7a, the device consists of an active chip with about 1500 microphotodiodes and an additional 4×4 array of light insensitive titanium nitride (TiN) electrodes (50 µm × 50 µm or 100 µm × 100 µm) for direct stimulation powered externally. Six patients, after being implanted with the device, could identify simple patterns, and another one was able to recognize complex letters and forms [68,69].

It is vital that visual stimulation electrodes should be placed close to the target neuron cells to achieve low threshold charge and high resolution. Palanker and co-workers compared different types (flat, pillars, and chambers) of passive subretinal arrays and found that pillars had minimal alteration of the inner retinal architecture as shown in Figure 7b [70]. They developed photovoltaic retinal prosthesis adopting optical amplification instead without the need for complex electrical circuitry and trans-scleral cabling [18,71]. Each element in the photodiode array consists of a central iridium oxide (IrO_x) electrode (Figure 7c). Photodiodes increase the dynamic range of the charge injection from the electrodes and maintain the light energy at safe levels [72]. An MIT-Harvard group developed Boston retinal implant which used a passive electrode array without MPDA for stimulation, and the new generation device had 256 electrodes as shown in Figure 7d. The earlier 15-channel prototype showed good tolerance and continued function after implanted in a mini-pig for 1 year [73].

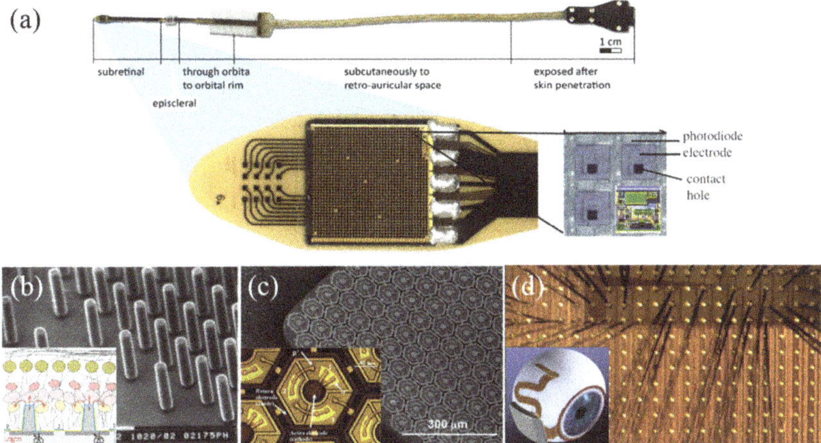

Figure 7. (**a**) The prototype of the Alpha-IMS predecessor, including 16 additional electrodes for direct stimulation. The microphotodiode array (MPDA) consists of 1500 photodiodes on a surface area of 3 × 3 mm [7]; (**b**) SEM of the microfabricated SU-8 pillar arrays, each pillar is about 10 μm in diameter and 40–70 μm in height. Insert: a pillar array attracts retinal cells for achieving intimate electrode-cell proximity [74]; (**c**) MPDA developed by the Palanker group. Insert: blown-up view of a single stimulating element with 3 photodiodes in series [7]; (**d**) Boston retinal implant chip, showing some of the 256 electrode current drivers. Insert: retinal implant concept with the secondary coil surrounding the cornea [73]. Reproduced from the mentioned references with permission from the related journals.

3.3. Supra-Choroidal Electrodes

The electrodes for supra-choroidal implant are relatively distant from the retina leading to higher stimulation threshold [75]. Kim and co-workers from Seoul National University built a prototype implant with 7-channel electrodes as shown in Figure 8a, where the reference electrode was placed on the outer scleral surface to simplify the surgical procedures and reduce the various risks of inserting electrodes into the vitreous [76].

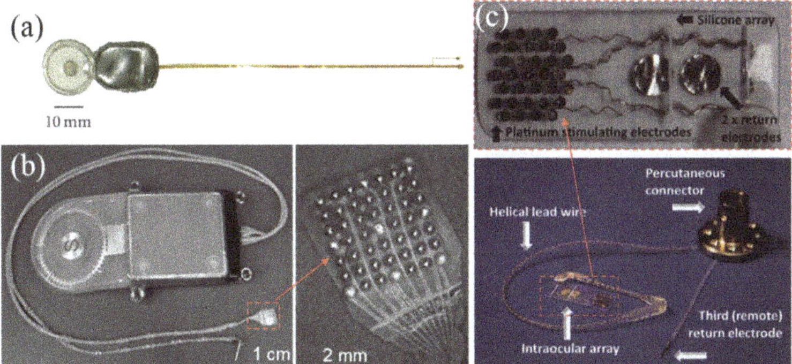

Figure 8. (**a**) A suprechoroidal implant for transretinal stimulation [70]; (**b**) the supra-choroidal-transretinal stimulation (STS) implant including the suprachoroidal stimulating array and the remote return electrode [77]; (**c**) the Bionic Vision Australia (BVA) implant with 1 remote return and 2 other return electrodes on the suprachoroidal array [15]. Reproduced from the mentioned references with permission from the related journals.

Fujikado et al. [78] developed a supra-choroidal-transretinal stimulation (STS) prosthesis as shown in Figure 8b, containing 49 Pt electrodes array with 500 μm in diameter and 700 μm separation, while only 9 electrodes were active. Nonetheless, it helped patients to localize objects better. Ayton et al. [15] from the Center for Eye Research Australia (CERA) developed another supra-choroidal prosthesis with the aim to achieve wide-view and high-resolution vision. As shown in Figure 8c, the intraocular array consists 33 Pt stimulating electrodes and 2 large return electrodes on silicone substrate. The implant remains stable during the Phase I clinical trial for 2 years [15].

4. Surface Modification for Electrodes

As the decreasing size of electrodes increases the impedance dramatically, it improves the signal-to-noise ratio (SNR) and reduces the stimulation efficiency in the clinical. Nanostructures and materials show great potential in improving the electrochemical and mechanical performances of neural interface in spite of the limitation size of electrodes. In the following sections, different coating materials including the key characterizations are reviewed in detail.

4.1. Metallic Materials and Their Derivatives

Noble metals such as platinum (Pt), gold (Au), and iridium (Ir) are the widely applied neural electrode materials for their excellent conductivity, stability, biocompatibility, and corrosion resistance [79,80]. Pt black coating was commonly formed by electrodeposition characterized with rough structure, which could be dated back to 1894 [81,82], reducing the impedance of microelectrode at least by a factor of 4 at 1 kHz [83,84]. However, its cytotoxicity remained a serious issue due to the lead (Pb) additive in electrolyte, which was strictly forbidden in the clinic trials [80].

In recent years, new methods have been developed to improve the neural interfaces. Pt gray was an alternative method without cytotoxic components patented by Second Sight [35,85,86], which was electrodeposited at an intermediate rate slower than that of Pt black and possessed desirable fractal morphology and stronger mechanical strength [79,86]. Pt gray had more than sufficient chronic charge injection capacity (CIC) up to 1.0 mC·cm^{-2} for retinal stimulation [66], nevertheless, its impedance and cathodic charge storage capacity (CSC$_c$) were still undesirable for higher density microelectrodes. To increase the electrochemical performances, Boehler et al. [80] deposited Pt-nanograss with large surface area, reducing the impedance by almost two orders of magnitude (at 1 kHz) and increasing the CSC$_c$ by ~40 times compared to bare electrode, shown in Figure 9a. In addition, it also exhibited good mechanical and electrochemical stability after cleaning and pulse testing [80]. Boretius et al. [87] introduced platinum–copper (Pt–Cu) alloys and then removed Cu to fabricate a cauliflower-like micro-sized Pt, shown in Figure 9b. It exhibited good electrochemical property due to its higher effective surface area. As shown in Figure 9c, other researchers applied templating to create well-ordered and micro-structured Pt coating with good mechanical property despite of its lower electrochemical performances than Pt black [88]. Pretreatment could also facilitate the roughening of coating by reactive ion etching (RIE) before sputtering Pt on bare electrode [89]. Our group has developed various well-controlled nanocrystal Pt coatings growing on the Pt electrode using different electrodeposition methods, which are featured with strong mechanical stability and ultra-high surface area and distinguished with randomly distributed nano-aggregation. A novel three-dimensional (3D) nanocrystal Pt coating shown in Figure 9d provided extremely large surface, which significantly reduced impedance of electrode by ~93% and increased its CSC$_c$ up to ~100 mC·cm^{-2}, showing good electrochemical stability with less than 3% loss after electrical stimulation for over 2×10^7 cycles [90].

Au was another promising material for neural electrode coating, the Au coating modified electrode developed by Kim et al. [91] displayed 4 times lower impedance. As shown in Figure 9e, the higher porous and interconnected Au coating obtained by co-depositing of gold–silver (Au–Ag) alloy and then removing Ag decreased the impedance by more than 25 times [92].

Despite of the advantages of Pt and Au, their CIC and CSC$_c$ were still limited with the shrinking size of the electrode sites for no faradaic reaction existed [93]. Thus, materials with a higher ability to support reversible faradaic reactions such as iridium oxide (IrO$_x$) were applied to promote the properties [36]. Considering the high cost and degradation/delaminate with repetitive stimulation, our group developed IrO$_x$/Pt nanocone composite coating (Figure 9f) by depositing a thin IrO$_x$ film on nanocone-shaped Pt to improve mechanical adhesion, demonstrating good CSC$_c$ of 22.29 mC·cm^{-2}, which was about 2.8 times higher than that of pure Pt coating [94]. And we further developed novel IrO$_x$/Pt nanoleaf composite coating by optimizing the deposition condition to get higher CSC$_c$ more than 400 mC·cm^{-2} with good stability (Figure 9g) [95], which is the best record so far.

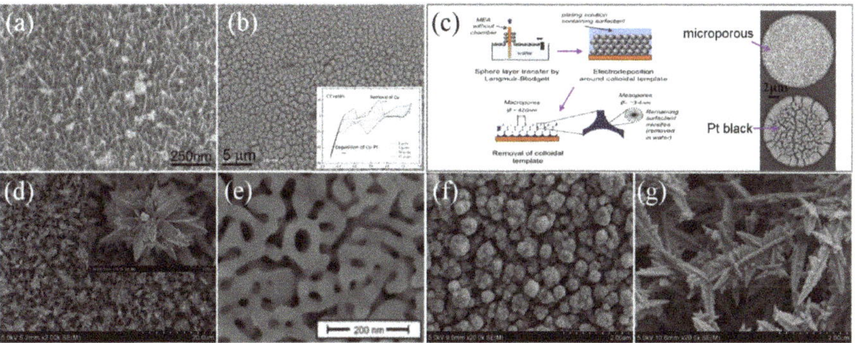

Figure 9. Scanning electron microscope (SEM) of different materials: (**a**) Pt-nanograss coating [80]; (**b**) cauliflower-like Pt coating [87]; (**c**) the fabrication process of depositing porous structure on electrode surface by templating [88]; (**d**) 3D nanocrystal Pt coating obtained by our group; (**e**) nanostructured Au coating by removing Ag from the Au–Ag alloy [96]; (**f**) IrO$_x$/Pt nanocone composite coating in our group [94]; (**g**) IrO$_x$/Pt nanoleaf composite coating in our group [95]. Reproduced from the mentioned references with permission from the related journals.

4.2. Conducting Polymers

Conducting polymers (CPs) are suitable candidate coating for neural interface which have been widely investigated, offering porous surfaces and pseudocapacitance, resulting in high CSC$_c$ and low impedance [97]. In recent years, most researches have been focused on poly (3,4-ethylenedioxythiophene) (PEDOT) for its higher conductivity and biocompatibility compared to other CPs such as polypyrrole (PPy) [98]. PEDOT had higher electroactivity and stability owing to its dioxyethylene bridging group that facilitated the charge transmit (Figure 10a) [99]. It had better electrochemical stability after stimulating with charge density up to 3 mC·cm^{-2}, compared to IrO$_x$ coating [100]. However, EDOT has low solubility in water and PEDOT has poor mechanical adhesion on the electrode because of its brittle structure [101,102]. Dopant can be introduced to improve the structural and properties of CP coatings, for example, polystyrene sulfonate (PSS) doped PEDOT displayed aggregation on the edge of the electrode while ClO$_4{}^-$ doped ones showed more even distribution [103]. Furthermore, the performances of CP coatings could be improved by means of structural design and surface functionalization. Abidian et al. [104] developed a 3D nanowire modified CP coating by electrospinning, demonstrating better elasticity and more porous morphology, which could help delivery drugs conveniently (Figure 10b). Templating was also used to create porous structures, the surface area could be increased by decreasing the size of polystyrene (PS) sphere, showing even lower impedance than bare electrode (Figure 10c), whereas the harmful solvents and residues in the templates needed to be further removed [105]. In addition, bioactive molecules could be added into the monomer of CPs in order to improve its biocompatibility and decrease the corresponding tissue response. Cui et al. [106] integrated Tyr-Ile-Gly-Ser-Arg (YIGSR) peptide into CPs to obtain

PPy/DCDPGYIGSR composite coating (Figure 10d) by electrodeposition. Their results showed that neural cells grew on 83% of the electrode sites with integrated peptides while only 10% on that without peptides. Richardson et al. [107] developed PPy/pTS/NT3 (Neurotrophin-3) which effectively avoided the degradation of auditory neurons and promoted the growth toward the electrode (Figure 10e). However, the long-term in vivo stability of CPs was still undesirable, which should be further studied in future since it remained a limiting factor for implantable electrodes.

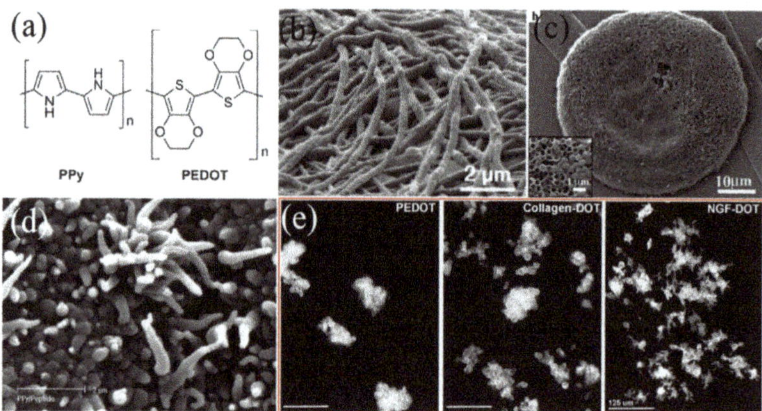

Figure 10. (**a**) Chemical structures of PPy and PEDOT [108]; (**b**) SEM image of PEDOT coating [109]; (**c**) SEM image of PEDOT obtained using polystyrene as template [105]; (**d**) SEM image of PEDOT/DCDPGYIGSR composite coating [106]; (**e**) Neurotrophic factor combined with PEDOT promoted the growth of neural cells [107]. Reproduced from the mentioned references with permission from the related journals.

4.3. Carbon Materials

Carbon materials are considered as another promising candidate since their toxicity is minimal compared to metallic materials [110]. Among them, carbon nanotube (CNT) shows the biggest advantage as neural electrode coating because of their high surface-to-volume ratio [111] and capacitive interfacial behavior during charge transfer [112]. Although the CIC of CNT was lower than that of IrO_x, it still showed safe stimulation in vitro [113]. CNT-coated electrode was found to be more sensitive to cell response than Pt electrode in vitro [114]. The properties of CNT was determined by its fabrication method. Chemical vapor deposition (CVD) was one of the most widely used technique, but the high temperature manufacturing process constrained the choice of electrode and substrate materials and needed additional transfer processes to polymer substrates [115]. Plasma treatment [115] and heat treatment [116] were adopted to remove amorphous carbon formed on the surface of CNTs during CVD process, which helped increasing 2-fold in capacitance but reduced the mechanical stability [116]. Furthermore, its high hydrophobicity significantly limited charge injection at the interface [113]. Thus, plasma treatment could be applied to form hydrophilic chemical bonds such as C–OH, C=O and OH–C=O, and the water contact angle could be also decreased by ultraviolet (UV)-ozone treatment (~145°) as shown in Figure 11a [115], leading to the increase of capacitance up to 80 folds. It could also promote the attachment and differentiation of neural cells to some degree *in vitro*, shown in Figure 11b [117]. Meanwhile, amination could further improve the electrochemical performance of CNT (Figure 11c) [118]. Almost no detachment occurred on the substrate after sonication, showing great adhesion strength (Figure 11d) [118]. However, their potential biotoxicity was still the main disadvantage.

In contrast, graphene bypassed this problem due to its planar geometry as well as good biocompatibility, leading to enhanced adhesion and viability [119]. Electrophoresis was usually employed to obtain graphene oxide (GO, Figure 11e) which exhibited high conductivity of 2000 S/m and low impedance, however, its smooth morphology limited its CSC_c lower than CNT [120,121]. The GO coating exhibited good cell viability more than 90% (Figure 11f), and it could reduce tissue responses with microglia and astrocytes distribution scattered around the coating in vitro study (Figure 11g) [122].

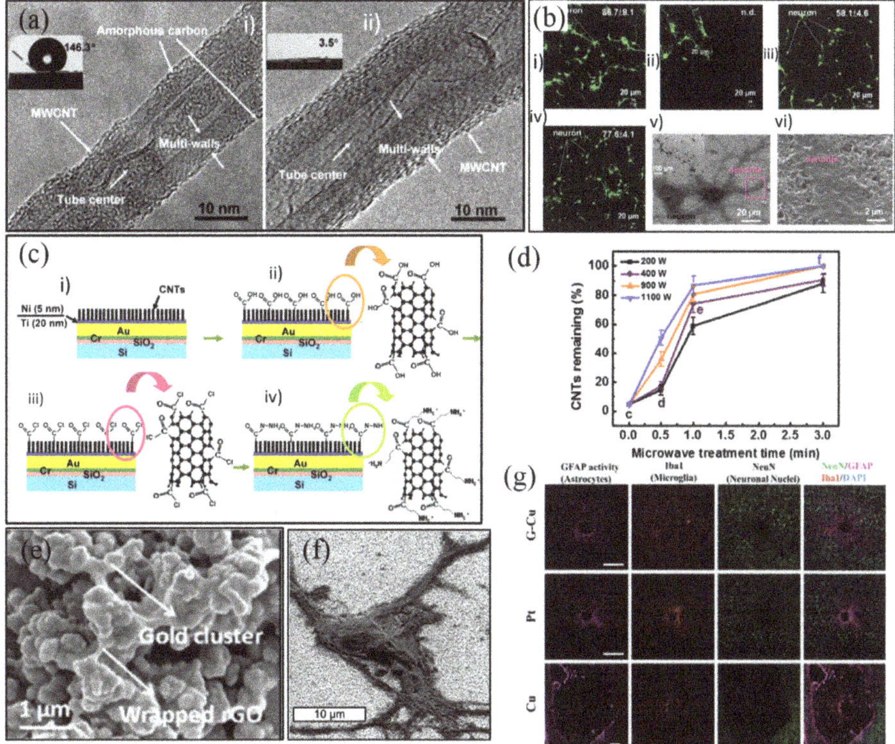

Figure 11. (**a**) Transmission electron microscope (TEM) of multi-walled carbon nanotube (MWCNT) with amorphous carbon before and after plasma treatment [115]; (**b**) fluorescent images of neuron cells cultured on as-grown carbon nanotubes (CNTs) and UV-ozone-modified CNTs, and corresponding SEM images [117]; (**c**) Process flow of MWCNT amino-functionalization [118]; (**d**) the percentages of CNTs remaining after 5 min sonication vs. microwave treatment time at various powers [123]; (**e**) SEM image of reduced graphene oxide (GO) coating [124]; (**f**) SEM images of neural cells attached well to GO coating [125]; (**g**) histological studies of tissue response to GO coating [122]. Reproduced from the mentioned references with permission from the related journals.

4.4. Composite Coatings

Although different materials have been studied for neural interfaces, no ideal material can perform well on all properties including electrical, mechanical, and biological. Hence, composites have been recently considered an attractive option to take advantages of several materials and bypass their disadvantages. Ferguson et al. [126] fabricated Au-CNT composite coating by combining the advantages of both materials, the mechanical adhesion was inherently improved as well as its surface area. The impedance of Au-CNT could be reduced by at least 10 times with much higher CSC_c than that of activated iridium oxide film (AIROF), while the CIC was less than 1 mC·cm^{-2}, since only a small amount of Faraday charge transferred between the tissue–electrode interface [126]. IrO_x could be used as an encapsulation layer for carbon bracket, which not only promoted more charge exchanging but also prevented carbon releasing [127]. Connecting IrO_x to CNT through the carboxylic acid groups improved the strength between them, leading to high effective surface area (Figure 12a) and much higher CSC_c of 101.2 mC·cm^{-2} than pure IrO_x (Figure 12b) [127]. Moreover, IrO_x could be combined with reduced graphene oxide (RGO) or GO to form composite coatings, among which IrO_x-GO showed larger CSC_c because of its rougher surface (Figure 12c) and remained more than 10% higher CSC_c than that of pure IrO_x and IrO_x-CNT after 1000 CV cycles, demonstrating great electrochemical stability [128,129].

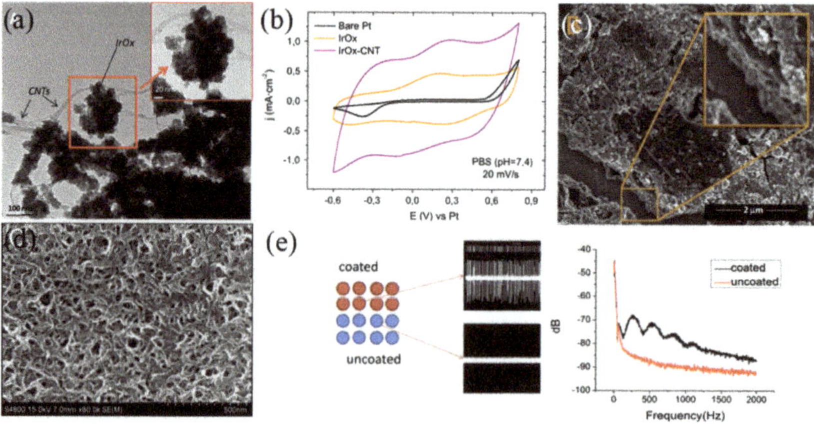

Figure 12. (**a**) TEM image of IrO_x-CNT composite coating [127]; (**b**) Cyclic voltammetry (CV) curves of bare Pt, IrO_x, and IrO_x-CNT coated electrodes in PBS (pH = 7.4) at sweep rate of 20 mV·s^{-1} [127]; (**c**) SEM image of IrO_x-GO composite coating [129]; (**d**) SEM image of SWNT-PPy composite coating [130]; (**e**) Influence of peptide bond on PEDOT/PSS/MWCNT composite coating [131]. Reproduced from the mentioned references with permission from the related journals.

In addition, combining CPs with nanostructured carbon materials (NC) like CNTs may also provide a highly electroactive, mechanically strong, and biocompatible coating. Zhou et al. [132] doped multi-walled CNT (MWCNT) in PEDOT to get PEDOT-MWCNT composite coating, exhibiting superior stability to pure PEDOT coating with only 2% loss of CSC_c after electrochemical stimulating for 96 h under 3 mC·cm^{-2} pulse. Compared to co-deposition of CP-NC coatings, electrodeposited CPs on NC layer by layer was more controllable (Figure 12d), the 3D independent topography could provide rougher surface to transfer charge with more electroactive points [133]. Hydrogels, biomolecules, etc. were often incorporated into the coating to further improve the biocompatibility, which was shown to benefit in vivo recording supported by more active channels and higher signal power than pure coating (Figure 12e) [131]. Anti-inflammatory drugs could also be incorporated in those composite coatings for alleviation [134].

5. Conclusions

This review summarizes the micro/nano technologies for high-density implantable packaging and fMEA as well as high-performance coating materials in the retinal prosthesis, which are crucial to achieve high resolution. Implantable biomedical component encapsulation is important, and high density, air tightness and biocompatible electronic packaging is an inevitable choice to build a sealed cavity with mechanical support performance and biological compatibility. Also, high-density microelectrode arrays should be capable of precise stimulation, while high-performance coating materials would help improve the implanted electrodes equipped with high CSC_c and CIC, and corresponding low impedance to ensure safe stimulation efficiently. It is worth considering that the chronic encapsulation, high-density fMEA together with high-performance coating materials contribute to high-density retinal implants, which lead to higher resolution, however, such hardware improvements need to be combined with better stimulation strategies and the latest progress in neuroscience and ophthalmology to realize the precise, effective, and chronic artificial vision.

Although the performance of retinal implants has been significantly improved through development in the past few decades, the design of the overall components is far from perfect, especially for high-density implants. There are still great challenges in high-density fabrication and integration, low invasiveness, power consumption, and high biocompatibility in the future. New concepts and materials are being introduced in the research of artificial vision, such as optogenetics [135], gene therapy [136], flexible photovoltaic films [137], and nanowire [138]. Future research efforts should target a more solid advancement in electronics, microfabrication, material science, and new biotechnologies in order to better understand retinal processing, so that the retinal implant will be easily accepted by the public with low cost.

Author Contributions: Conceptualization, T.W. and Y.Z.; Chapter on package, S.Z and H.Y.; Chapter on electrode, Q.Z.; Chapter on coating, Q.Z.; Introduction and summary, T.W., Q.Z. and S.Z.; Review—T.W., Y.Z.; Revision: Q.Z., S.Z., T.W.

Funding: This research was funded by CAS STS program (grant number KFJ-STS-SCYD-225); Guangdong Natural Foundation (grant number 2015A030306021); Shenzhen Peacock Plan (grant number 20130409162728468); Shenzhen Science and Technology Research Program (grant number JCYJ20170818152810899, JCYJ20170818154035069, JCYJ201606112152651093) and Shenzhen Maker and Start-up Funding Project (grant number CKCY2018032911060178).

Acknowledgments: Authors acknowledged Bo Peng, Fengqi Yu and Chunlei Yang at SIAT for their kind help and contribution in the R&D of the SIAT retinal implant with 100+ and 1000+ channels.

References

1. Schwarz, D.A.; Lebedev, M.A.; Hanson, T.L.; Dimitrov, D.F.; Lehew, G.; Meloy, J.; Rajangam, S.; Subramanian, V.; Ifft, P.J.; Li, Z. Chronic, wireless recordings of large-scale brain activity in freely moving rhesus monkeys. *Nat. Methods* **2014**, *11*, 670. [CrossRef] [PubMed]

2. Lee, S.B.; Yin, M.; Manns, J.R.; Ghovanloo, M. A wideband dual-antenna receiver for wireless recording from animals behaving in large arenas. *IEEE Trans. Biomed. Eng.* **2013**, *60*, 1993–2004. [PubMed]

3. Yin, M.; Borton, D.A.; Komar, J.; Agha, N.; Lu, Y.; Li, H.; Laurens, J.; Lang, Y.; Li, Q.; Bull, C. Wireless neurosensor for full-spectrum electrophysiology recordings during free behavior. *Neuron* **2014**, *84*, 1170–1182. [CrossRef]

4. Greatbatch, W.; Holmes, C.F. History of implantable devices. *IEEE Eng. Med. Biol. Mag.* **1991**, *10*, 38–41. [CrossRef] [PubMed]

5. Magjarević, R.; Ferek-Petrić, B. Implantable cardiac pacemakers–50 years from the first implantation. *Slov. Med. J.* **2010**, *79*, 55–67.

6. Lee, S.; Koo, K.-i.; Ko, H.; Seo, J.-M.; Cho, D.-i.D. Review of high-resolution retinal prosthetic system for vision rehabilitation: Our perspective based on 18 years of research. *Sens. Mater.* **2018**, *30*, 1393–1406. [CrossRef]

7. Yue, L.; Weiland, J.D.; Roska, B.; Humayun, M.S. Retinal stimulation strategies to restore vision: Fundamentals and systems. *Prog. Retin. Eye Res.* **2016**, *53*, 21–47. [CrossRef]

8. Rizzo, S.; Barale, P.O.; Ayello-Scheer, S.; Devenyi, R.G.; Delyfer, M.N.; Korobelnik, J.F.; Rachitskaya, A.; Yuan, A.; Jayasundera, K.T.; Zacks, D.N.; et al. Adverse events of the Argus II retinal prosthesis: Incidence, causes, and best practices for managing and preventing conjunctival erosion. *Retina* **2018**, *0*, 1–9. [CrossRef]

9. Luo, H.L.; Cruz, L.D. The Argus ®; II retinal prosthesis system. *Prog. Retin. Eye Res.* **2016**, *50*, 89–107. [CrossRef]

10. Fernandes, R.A.B.; Diniz, B.; Ribeiro, R.; Humayun, M. Artificial vision through neuronal stimulation. *Neurosci. Lett.* **2012**, *519*, 122–128. [CrossRef]

11. Hadjinicolaou, A.E.; Meffin, H.; Maturana, M.I.; Cloherty, S.L.; Ibbotson, M.R. Prosthetic vision: Devices, patient outcomes and retinal research. *Clin. Exp. Optom.* **2015**, *98*, 395–410. [CrossRef] [PubMed]

12. Luo, H.L.; Fukushige, E.; Cruz, L.D. The potential of the second sight system bionic eye implant for partial sight restoration. *Expert Rev. Med. Devices* **2016**, *13*, 673–681. [CrossRef] [PubMed]

13. Hafed, Z.M.; Stingl, K.; Bartz-Schmidt, K.-U.; Gekeler, F.; Zrenner, E. Oculomotor behavior of blind patients seeing with a subretinal visual implant. *Vis. Res.* **2016**, *118*, 119–131. [CrossRef] [PubMed]

14. Fujikado, T.; Kamei, M.; Sakaguchi, H.; Kanda, H.; Endo, T.; Hirota, M.; Morimoto, T.; Nishida, K.; Kishima, H.; Terasawa, Y. One-year outcome of 49-channel suprachoroidal-transretinal stimulation prosthesis in patients with advanced retinitis pigmentosa. *Investig. Ophthalmol. Vis. Sci.* **2016**, *57*, 6147–6157. [CrossRef] [PubMed]

15. Ayton, L.N.; Blamey, P.J.; Guymer, R.H.; Luu, C.D.; Nayagam, D.A.; Sinclair, N.C.; Shivdasani, M.N.; Yeoh, J.; Mccombe, M.F.; Briggs, R.J. First-in-human trial of a novel suprachoroidal retinal prosthesis. *PLoS ONE* **2014**, *9*, e115239. [CrossRef] [PubMed]

16. Zrenner, E. Fighting blindness with microelectronics. *Sci. Transl. Med.* **2013**, *5*, 210ps216. [CrossRef]

17. Coffey, V.C. Vision accomplished: the bionic eye. *Opt. Photonics News* **2017**, *28*, 24–31. [CrossRef]

18. Palanker, D.; Vankov, A.; Huie, P.; Baccus, S. Design of a high-resolution optoelectronic retinal prosthesis. *J. Neural. Eng.* **2005**, *2*, S105. [CrossRef]

19. Kang, H.; Abbasi, W.H.; Kim, S.W.; Kim, J. Fully integrated light-sensing stimulator fesign for dubretinal implants. *Sensors* **2019**, *19*, E536. [CrossRef]

20. Weiland, J.D.; Humayun, M.S. Visual prosthesis. *Proc. IEEE* **2008**, *96*, 1076–1084. [CrossRef]

21. Franz, S.; Rammelt, S.; Scharnweber, D.; Simon, J.C. Immune responses to implants—A review of the implications for the design of immunomodulatory biomaterials. *Biomaterials* **2011**, *32*, 6692–6709. [CrossRef] [PubMed]

22. Williams, D.F. On the mechanisms of biocompatibility. *Biomaterials* **2008**, *29*, 2941–2953. [CrossRef] [PubMed]

23. Polikov, V.S.; Tresco, P.A.; Reichert, W.M. Response of brain tissue to chronically implanted neural electrodes. *J. Neurosci. Methods* **2005**, *148*, 1–18. [CrossRef] [PubMed]

24. Lichter, S.G.; Escudié, M.C.; Stacey, A.D.; Ganesan, K.; Fox, K.; Ahnood, A.; Apollo, N.V.; Kua, D.C.; Lee, A.Z.; McGowan, C. Hermetic diamond capsules for biomedical implants enabled by gold active braze alloys. *Biomaterials* **2015**, *53*, 464–474. [CrossRef] [PubMed]

25. Ganesan, K.; Garrett, D.J.; Ahnood, A.; Shivdasani, M.N.; Tong, W.; Turnley, A.M.; Fox, K.; Meffin, H.; Prawer, S. An all-diamond, hermetic electrical feedthrough array for a retinal prosthesis. *Biomaterials* **2014**, *35*, 908–915. [CrossRef] [PubMed]

26. Sundaram, V.; Sukumaran, V.; Cato, M.E.; Liu, F.; Tummala, R.; Weiland, J.D.; Nasiatka, P.J.; Tanguay, A.R. High density electrical interconnections in liquid crystal polymer (LCP) substrates for retinal and neural prosthesis applications. In Proceedings of the 61st IEEE Electronic Components and Technology Conference (ECTC), Lake Buena Vista, FL, USA, 31 May–1 June 2011; pp. 1308–1313.

27. Schuettler, M.; Ordonez, J.S.; Santisteban, T.S.; Schatz, A.; Wilde, J.; Stieglitz, T. Fabrication and test of a hermetic miniature implant package with 360 electrical feedthroughs. In Proceedings of the 32nd Annual International Conference of the IEEE Engineering in Medicine and Biology Society (EMBC 2010), Buenos Aires, Argentina, 31 August–4 September 2010; pp. 1585–1588.

28. Suaning, G.J.; Lavoie, P.; Forrester, J.; Armitage, T.; Lovell, N.H. Microelectronic retinal prosthesis: III. A new method for fabrication of high-density hermetic feedthroughs. In Proceedings of 28th Annual International Conference of the IEEE Engineering in Medicine and Biology Society (EMBC 2006), New York, NY, USA, 30 August–3 September 2006; pp. 1638–1641.

29. Carnicerlombarte, A.; Lancashire, H.; Vanhoestenberghe, A. In vitro biocompatibility and electrical stability of thick-film platinum/gold alloy electrodes printed on alumina. *J. Neural Eng.* **2017**, *14*, 036012. [CrossRef]

30. Deng, M.; Yang, X.; Silke, M.; Qiu, W.; Xu, M.; Borghs, G.; Chen, H. Electrochemical deposition of polypyrrole/graphene oxide composite on microelectrodes towards tuning the electrochemical properties of neural probes. *Sens. Actuators B Chem.* **2011**, *158*, 176–184. [CrossRef]

31. Velliste, M.; Perel, S.; Spalding, M.C.; Whitford, A.S.; Schwartz, A.B. Cortical control of a prosthetic arm for self-feeding. *Nature* **2008**, *453*, 1098–1101. [CrossRef]

32. Sekirnjak, C.; Hottowy, P.; Sher, A.; Dabrowski, W.; Litke, A.; Chichilnisky, E. Electrical stimulation of mammalian retinal ganglion cells with multielectrode arrays. *J. Neurophysiol.* **2006**, *95*, 3311–3327. [CrossRef]

33. Rodger, D.C.; Fong, A.J.; Li, W.; Ameri, H.; Ahuja, A.K.; Gutierrez, C.; Lavrov, I.; Zhong, H.; Menon, P.R.; Meng, E. Flexible parylene-based multielectrode array technology for high-density neural stimulation and recording. *Sens. Actuators B Chem.* **2008**, *132*, 449–460. [CrossRef]

34. Negi, S.; Bhandari, R.; Rieth, L.; Van Wagenen, R.; Solzbacher, F. Neural electrode degradation from continuous electrical stimulation: Comparison of sputtered and activated iridium oxide. *J. Neurosci. Methods* **2010**, *186*, 8–17. [CrossRef] [PubMed]

35. Zhou, D.D.; Talbot, N.H.; Greenberg, R.J. Adherent Metal Oxide Coating Forming a High Surface Area Electrode. U.S. Patent 9308299B2, 12 April 2016.

36. Lu, Y.; Wang, T.; Cai, Z.; Cao, Y.; Yang, H.; Duan, Y.Y. Anodically electrodeposited iridium oxide films microelectrodes for neural microstimulation and recording. *Sens. Actuators B Chem.* **2009**, *137*, 334–339. [CrossRef]

37. Picaud, S.; Sahel, J.-A. Retinal prostheses: Clinical results and future challenges. *C. R. Biol.* **2014**, *337*, 214–222. [CrossRef] [PubMed]

38. Chuang, A.T.; Margo, C.E.; Greenberg, P.B. Retinal implants: a systematic review. *Br. J. Ophthalmol.* **2014**, *98*, 852–856. [CrossRef] [PubMed]

39. Diego, G. Retinal prostheses: Progress toward the next generation implants. *Front. Neurosci.* **2015**, *9*, 290.

40. Wen, H.K. Early history and challenges of implantable electronics. *ACM J. Emerg. Technol. Comput. Syst.* **2012**, *8*, 1–9.

41. Teo, A.J.T.; Mishra, A.; Park, I.; Kim, Y.J.; Yong, J.Y. Polymeric biomaterials for medical implants & devices. *ACS Biomater. Sci. Eng.* **2016**, *2*. [CrossRef]

42. Ok, J.; Greenberg, R.J. Method and Apparatus for Providing Hermetic Electrical Feedthrough. U.S. Patent 8163397B2, 24 April 2012.

43. Yang, H.; Wu, T.; Zhao, S.; Xiong, S.; Peng, B.; Humayun, M.S. Chronically implantable package based on alumina ceramics and titanium with high-density feedthroughs for medical implants. In Proceedings of the 40th 2018 Annual International Conference of the IEEE Engineering in Medicine and Biology Society (EMBC), Honolulu, HI, USA, 17–21 July 2018; pp. 3382–3385.

44. Guenther, T.; Kong, C.; Lu, H.; Svehla, M.J.; Lovell, N.H.; Ruys, A.; Suaning, G.J. Pt-Al$_2$O$_3$ interfaces in cofired ceramics for use in miniaturized neuroprosthetic implants. *J. Biomed. Mater. Res. Part B Appl. Biomater.* **2014**, *102*, 500–507. [CrossRef]

45. Green, R.A.; Thomas, G.; Christoph, J.; Amandine, J.; Yu, J.F.; Dueck, W.F.; Lim, W.W.; Henderson, W.C.; Anne, V.; Lovell, N.H.; et al. Integrated electrode and high density feedthrough system for chip-scale implantable devices. *Biomaterials* **2013**, *34*, 6109–6118. [CrossRef]

46. Nomura, K.; Okada, A.; Shoji, S.; Ogashiwa, T.; Mizuno, J. Application of I-structure though-glass interconnect filled with submicron gold particles to a hermetic sealing device. *J. Micromech. Microeng.* **2016**, *26*, 105018. [CrossRef]

47. Brabec, C.; Dyakonov, V.; Scherf, U. *Organic Photovoltaics: Materials, Device Physics, and Manufacturing Technologies*; Wiley-VCH: Weinheim, Germany, 2009.

48. Stark, N. Literature review: Biological safety of parylene C. *Med Plast. Biomater.* **1996**, *3*, 30–35.

49. Li, W.; Rodger, D.C.; Pinto, A.; Meng, E.; Weiland, J.D.; Humayun, M.S.; Tai, Y.-C. Parylene-based integrated wireless single-channel neurostimulator. *Sens. Actuators A Phys.* **2011**, *166*, 193–200. [CrossRef]

50. Zhao, S.; Feng, Y.; Wu, T.; Yang, C.Y. Application of biocompatible organic/inorganic composite film in implantable retinal prosthesis. In Proceedings of the IEEE International Conference on Cyborg and Bionic Systems (CBS 2018), Shenzhen, China, 25–27 October 2018; pp. 101–105.

51. Zhao, S.; Feng, Y.; Xiong, S.; Peng, B.; Sun, B.; Wu, T.; Yang, C. Biocompatible organic/inorganic composite film packaging for implantable retinal prosthesis. In Proceedings of the 13th IEEE Annual International Conference on Nano/Micro Engineered and Molecular Systems (NEMS 2018), Singapore, 22–26 October 2018; pp. 325–328.

52. Jeong, J.; Lee, S.W.; Min, K.S.; Kim, S.J. A novel multilayered planar coil based on biocompatible liquid crystal polymer for chronic implantation. *Sens. Actuators A Phys.* **2013**, *197*, 38–46. [CrossRef]

53. Gwon, T.M.; Kim, C.; Shin, S.; Park, J.H.; Jin, H.K.; Kim, S.J. Liquid crystal polymer (LCP)-based neural prosthetic devices. *Biomed. Eng. Lett.* **2016**, *6*, 148–163. [CrossRef]

54. Dutta, D.; Fruitwala, H.; Kohli, A.; Weiss, R.A. Polymer blends containing liquid crystals: A Review. *Polym. Eng. Sci.* **2010**, *30*, 1005–1018. [CrossRef]

55. Joonsoo, J.; So, H.B.; Kyou, S.M.; Jong-Mo, S.; Hum, C.; Sung, J.K. A miniaturized, eye-conformable, and long-term reliable retinal prosthesis using monolithic fabrication of liquid crystal polymer (LCP). *IEEE Trans. Biomed Eng.* **2015**, *62*, 982–989.

56. Robblee, L.S.; Rose, T.L. Electrochemical guidelines for selection of protocols and electrode materials for neural stimulation. *Neural Prostheses Fundam. Stud.* **1990**, *1*, 25–66.

57. Humayun, M.S.; De Juan, E.; Dagnelie, G.; Greenberg, R.J.; Propst, R.H.; Phillips, D.H. Visual perception elicited by electrical stimulation of retina in blind humans. *Arch. Ophthalmol.* **1996**, *114*, 40–46. [CrossRef]

58. Caspi, A.; Dorn, J.D.; McClure, K.H.; Humayun, M.S.; Greenberg, R.J.; McMahon, M.J. Feasibility study of a retinal prosthesis: spatial vision with a 16-electrode implant. *Arch. Ophthalmol.* **2009**, *127*, 398–401. [CrossRef]

59. Hornig, R.; Zehnder, T.; Velikay-Parel, M.; Laube, T.; Feucht, M.; Richard, G. The IMI retinal implant system. In *Artificial Sight*; Humayun, M.S., Weiland, J.D., Chader, G., Greenbaum, E., Eds.; Springer: Basel, Switzerland, 2007; pp. 111–128.

60. Susanne, K.; Michael, G.; Stefan, R.; Dirk, H.; Uwe, T.; Reinhard, E.; Frank, B.; Thomas, W. Stimulation with a wireless intraocular epiretinal implant elicits visual percepts in blind humans. *Investig. Ophthalmol. Vis. Sci.* **2011**, *52*, 449–455.

61. Edwards, T.L.; Cottriall, C.L.; Xue, K.; Simunovic, M.P.; Ramsden, J.D.; Zrenner, E.; MacLaren, R.E. Assessment of the electronic retinal implant alpha AMS in restoring vision to blind patients with end-stage retinitis pigmentosa. *Ophthalmology* **2018**, *125*, 432–443. [CrossRef] [PubMed]

62. Daschner, R.; Rothermel, A.; Rudorf, R.; Rudorf, S.; Stett, A. Functionality and performance of the subretinal implant chip Alpha AMS. *Sens. Mater.* **2018**, *30*, 179–192. [CrossRef]

63. Xia, K.; Sun, B.; Zeng, Q.; Wu, T.; Humayun, M.S. Surface Modification of Neural Stimulating/Recording Microelectrodes with High-Performance Platinum-Pillar Coatings. In Proceedings of the 12th IEEE International Conference on Nano/Micro Engineered and Molecular Systems (NEMS 2017), Los Angeles, CA, USA, 9–12 April 2017; pp. 291–294.

64. Sun, B.; Li, T.; Xia, K.; Zeng, Q.; Wu, T.; Humayun, M.S. Flexible Microelectrode Array for Retinal Prosthesis. In Proceedings of the 39th Annual International Conference of the IEEE Engineering in Medicine and Biology Society (EMBC 2017), Jeju Island, Korea, 11–15 July 2017; pp. 1097–1100.

65. Li, T.; Sun, B.; Xia, K.; Zeng, Q.; Wu, T.; Humayun, M.S. Design and fabrication of a high-density flexible microelectrode array. Proceedings of 12th IEEE International Conference on Nano/Micro Engineered and Molecular Systems (NEMS 2017), Los Angeles, CA, USA, 9–12 April 2017; pp. 299–302.

66. Zhou, D.D.; Greenbaum, E. *Implantable Neural Prostheses 1*; Biological and Medical Physics, Biomedical Engineering Book Series; Springer: Basel, Switzerland, 2009.

67. Chow, A.Y.; Bittner, A.K.; Pardue, M.T. The artificial silicon retina in retinitis pigmentosa patients (an American Ophthalmological Association thesis). *Trans. Am. Ophthalmol. Soc.* **2010**, *108*, 120. [PubMed]

68. Wilke, R.; Gabel, V.-P.; Sachs, H.; Schmidt, K.-U.B.; Gekeler, F.; Besch, D.; Szurman, P.; Stett, A.; Wilhelm, B.; Peters, T. Spatial resolution and perception of patterns mediated by a subretinal 16-electrode array in patients blinded by hereditary retinal dystrophies. *Investig. Ophthalmol. Vis. Sci.* **2011**, *52*, 5995–6003. [CrossRef] [PubMed]

69. Zrenner, E.; Bartz-Schmidt, K.U.; Benav, H.; Besch, D.; Bruckmann, A.; Gabel, V.P.; Gekeler, F.; Greppmaier, U.; Harscher, A.; Kibbel, S.; et al. Subretinal electronic chips allow blind patients to read letters and combine them to words. *Proc. Biol. Sci.* **2011**, *278*, 1489–1497. [CrossRef] [PubMed]

70. Loudin, J.; Simanovskii, D.; Vijayraghavan, K.; Sramek, C.; Butterwick, A.; Huie, P.; McLean, G.; Palanker, D. Optoelectronic retinal prosthesis: system design and performance. *J. Neural Eng.* **2007**, *4*, S72. [CrossRef] [PubMed]

71. Mathieson, K.; Loudin, J.; Goetz, G.; Huie, P.; Wang, L.; Kamins, T.I.; Galambos, L.; Smith, R.; Harris, J.S.; Sher, A. Photovoltaic retinal prosthesis with high pixel density. *Nat. Photonics* **2012**, *6*, 391. [CrossRef]

72. Wang, L.; Mathieson, K.; Kamins, T.I.; Loudin, J.D.; Galambos, L.; Goetz, G.; Sher, A.; Mandel, Y.; Huie, P.; Lavinsky, D.; et al. Photovoltaic retinal prosthesis: implant fabrication and performance. *J. Neural Eng.* **2012**, *9*, 046014. [CrossRef]

73. Kelly, S.K.; Shire, D.B.; Chen, J.; Gingerich, M.D.; Cogan, S.F.; Drohan, W.A.; Ellersick, W.; Krishnan, A.; Behan, S.; Wyatt, J.L. Developments on the Boston 256-channel retinal implant. In Proceedings of the 2013 IEEE International Conference on Multimedia and Expo Workshops (ICMEW), San Jose, CA, USA, 15–19 July 2013; pp. 1–6.

74. Zhou, D.D.; Greenberg, R.J. *Microelectronic Visual Prostheses*; Springer: Basel, Switzerland, 2009.

75. Yamauchi, Y.; Franco, L.M.; Jackson, D.J.; Naber, J.F.; Ziv, R.O.; Rizzo, J.F., III; Kaplan, H.J.; Enzmann, V. Comparison of electrically evoked cortical potential thresholds generated with subretinal or suprachoroidal placement of a microelectrode array in the rabbit. *J. Neural Eng.* **2005**, *2*, S48. [CrossRef]

76. Sakaguchi, H.; Fujikado, T.; Fang, X.; Kanda, H.; Osanai, M.; Nakauchi, K.; Ikuno, Y.; Kamei, M.; Yagi, T.; Nishimura, S. Transretinal electrical stimulation with a suprachoroidal multichannel electrode in rabbit eyes. *Jpn. J. Ophthalmol.* **2004**, *48*, 256–261. [CrossRef]

77. Morimoto, T.; Kamei, M.; Nishida, K.; Sakaguchi, H.; Kanda, H.; Ikuno, Y.; Kishima, H.; Maruo, T.; Konoma, K.; Ozawa, M.; et al. Chronic implantation of newly developed suprachoroidal-transretinal stimulation prosthesis in dogs. *Investig. Ophthalmol. Vis. Sci.* **2011**, *52*, 6785–6792. [CrossRef] [PubMed]

78. Fujikado, T.; Kamei, M.; Sakaguchi, H.; Kanda, H.; Morimoto, T.; Ikuno, Y.; Nishida, K.; Kishima, H.; Maruo, T.; Konoma, K.; et al. Testing of semichronically implanted retinal prosthesis by suprachoroidal-transretinal stimulation in patients with retinitis pigmentosa. *Investig. Ophthalmol. Vis. Sci.* **2011**, *52*, 4726–4733. [CrossRef] [PubMed]

79. Cogan, S.F. Neural stimulation and recording electrodes. *Annu. Rev. Biomed. Eng.* **2008**, *10*, 275–309. [CrossRef] [PubMed]

80. Boehler, C.; Stieglitz, T.; Asplund, M. Nanostructured platinum grass enables superior impedance reduction for neural microelectrodes. *Biomaterials* **2015**, *67*, 346–353. [CrossRef] [PubMed]

81. Marrese, C.A. Preparation of strongly adherent platinum black coatings. *Anal. Chem.* **1987**, *59*, 217–218. [CrossRef]

82. Lummer, O.; Kurlbaum, F. About the new platinum light unit. *Proc. Ger. Phys. Soc. Berl.* **1895**, *14*, 56–70.

83. Borkholder, D.; Bao, J.; Maluf, N.; Perl, E.; Kovacs, G. Microelectrode arrays for stimulation of neural slice preparations. *J. Neurosci. Methods* **1997**, *77*, 61–66. [CrossRef]

84. Arcot Desai, S.; Rolston, J.D.; Guo, L.; Potter, S.M. Improving impedance of implantable microwire multi-electrode arrays by ultrasonic electroplating of durable platinum black. *Front. Neuroeng.* **2010**, *3*, 5.

85. Zhou, D.M.; Ok, J.; Talbot, N.H.; Mech, B.V.; Little, J.S.; Greenberg, R.J. Electrode with Increased Stability and Method of Manufacturing the Same. U.S. Patent 7937153B2, 3 May 2011.

86. Zhou, D.M. Platinum Electrode Surface Coating and Method for Manufacturing the Same. U.S. Patent 7887687B2, 15 February 2011.

87. Boretius, T.; Jurzinsky, T.; Koehler, C.; Kerzenmacher, S.; Hillebrecht, H.; Stieglitz, T. High-porous platinum electrodes for functional electrical stimulation. In Proceedings of the 2011 Annual International Conference of the IEEE Engineering in Medicine and Biology Society, (EMBC), Boston, MA, USA, 30 August–3 September 2011; pp. 5404–5407.

88. Heim, M.; Rousseau, L.; Reculusa, S.; Urbanova, V.; Mazzocco, C.; Joucla, S.; Bouffier, L.; Vytras, K.; Bartlett, P.; Kuhn, A. Combined macro-/mesoporous microelectrode arrays for low-noise extracellular recording of neural networks. *J. Neurophysiol.* **2012**, *108*, 1793–1803. [CrossRef]

89. Negi, S.; Bhandari, R.; Solzbacher, F. A novel technique for increasing charge injection capacity of neural electrodes for efficacious and safe neural stimulation. In Proceedings of the 2012 Annual International Conference of the IEEE Engineering in Medicine and Biology Society (EMBC), San Diego, CA, USA, 28 August–1 September 2012; pp. 5142–5145.

90. Zeng, Q.; Zhang, Y.; Wu, T.; Sun, B.; Xia, K.; Humayun, M.S. 3D nano-crystal platinum for high-performance neural electrode. In Proceedings of the 40th Annual International Conference of the IEEE Engineering in Medicine and Biology Society (EMBC 2018), Honolulu, HI, USA, 17–21 July 2018; pp. 4217–4220.

91. Kim, M.-W.; Song, Y.-H.; Yang, H.-H.; Yoon, J.-B. An ultra-low voltage MEMS switch using stiction-recovery actuation. *J. Micromech. Microeng.* **2013**, *23*, 045022. [CrossRef]

92. Seker, E.; Berdichevsky, Y.; Begley, M.R.; Reed, M.L.; Staley, K.J.; Yarmush, M.L. The fabrication of low-impedance nanoporous gold multiple-electrode arrays for neural electrophysiology studies. *Nanotechnology* **2010**, *21*, 125504. [CrossRef] [PubMed]

93. Yoo, J.-M.; Negi, S.; Tathireddy, P.; Solzbacher, F.; Song, J.-I.; Rieth, L.W. Excimer laser deinsulation of Parylene-C on iridium for use in an activated iridium oxide film-coated Utah electrode array. *J. Neurosci. Methods* **2013**, *215*, 78–87. [CrossRef] [PubMed]

94. Zeng, Q.; Xia, K.; Sun, B.; Yin, Y.; Wu, T.; Humayun, M.S. Electrodeposited iridium oxide on platinum nanocones for improving neural stimulation microelectrodes. *Electrochim. Acta* **2017**, *237*, 152–159. [CrossRef]

95. Zeng, Q.; Xia, K.; Sun, B.; Wu, T.; Humayun, M.S. High-performance iridium oxide/platinum nano-leaf composite coatings on microelectrodes for neural stimulation/recording. In Proceedings of the 39th Annual International Conference of the IEEE Engineering in Medicine and Biology Society (EMBC 2017), Jeju Island, Korea, 11–15 July 2017; pp. 1070–1073.

96. Chapman, C.A.; Chen, H.; Stamou, M.; Biener, J.; Biener, M.M.; Lein, P.J.; Seker, E. Nanoporous gold as a neural interface coating: effects of topography, surface chemistry, and feature size. *ACS Appl. Mater. Interfaces* **2015**, *7*, 7093–7100. [CrossRef] [PubMed]

97. Rivnay, J.; Inal, S.; Collins, B.A.; Sessolo, M.; Stavrinidou, E.; Strakosas, X.; Tassone, C.; Delongchamp, D.M.; Malliaras, G.G. Structural control of mixed ionic and electronic transport in conducting polymers. *Nat. Commun.* **2016**, *7*, 11287. [CrossRef] [PubMed]

98. Ludwig, K.A.; Uram, J.D.; Yang, J.; Martin, D.C.; Kipke, D.R. Chronic neural recordings using silicon microelectrode arrays electrochemically deposited with a poly (3, 4-ethylenedioxythiophene) (PEDOT) film. *J. Neural Eng.* **2006**, *3*, 59. [CrossRef]

99. Che, J.; Xiao, Y.; Zhu, X.; Sun, X. Electro-synthesized PEDOT/glutamate chemically modified electrode: a combination of electrical and biocompatible features. *Polym. Int.* **2008**, *57*, 750–755. [CrossRef]

100. Venkatraman, S.; Hendricks, J.; King, Z.A.; Sereno, A.J.; Richardson-Burns, S.; Martin, D.; Carmena, J.M. In vitro and in vivo evaluation of PEDOT microelectrodes for neural stimulation and recording. *IEEE Trans. Neural Syst. Rehabil. Eng.* **2011**, *19*, 307–316. [CrossRef]

101. Abidian, M.R.; Corey, J.M.; Kipke, D.R.; Martin, D.C. Conducting-polymer nanotubes improve electrical properties, mechanical adhesion, neural attachment, and neurite outgrowth of neural electrodes. *Small* **2010**, *6*, 421–429. [CrossRef]

102. Cui, X.; Martin, D.C. Electrochemical deposition and characterization of poly (3, 4-ethylenedioxythiophene) on neural microelectrode arrays. *Sens. Actuators B Chem.* **2003**, *89*, 92–102. [CrossRef]

103. Green, R.A.; Hassarati, R.T.; Bouchinet, L.; Lee, C.S.; Cheong, G.L.; Jin, F.Y.; Dodds, C.W.; Suaning, G.J.; Poole-Warren, L.A.; Lovell, N.H. Substrate dependent stability of conducting polymer coatings on medical electrodes. *Biomaterials* **2012**, *33*, 5875–5886. [CrossRef] [PubMed]

104. Abidian, M.R.; Martin, D.C. Experimental and theoretical characterization of implantable neural microelectrodes modified with conducting polymer nanotubes. *Biomaterials* **2008**, *29*, 1273–1283. [CrossRef] [PubMed]

105. Yang, J.; Martin, D.C. Microporous conducting polymers on neural microelectrode arrays: I Electrochemical deposition. *Sens. Actuators B Chem.* **2004**, *101*, 133–142. [CrossRef]

106. Cui, X.; Wiler, J.; Dzaman, M.; Altschuler, R.A.; Martin, D.C. In vivo studies of polypyrrole/peptide coated neural probes. *Biomaterials* **2003**, *24*, 777–787. [CrossRef]

107. Kim, D.H.; Richardson-Burns, S.M.; Hendricks, J.L.; Sequera, C.; Martin, D.C. Effect of immobilized nerve growth factor on conductive polymers: Electrical properties and cellular response. *Adv. Funct. Mater.* **2007**, *17*, 79–86. [CrossRef]

108. Green, R.; Abidian, M.R. Conducting polymers for neural prosthetic and neural interface applications. *Adv. Mater.* **2015**, *27*, 7620–7637. [CrossRef] [PubMed]

109. Abidian, M.R.; Kim, D.H.; Martin, D.C. Conducting-polymer nanotubes for controlled drug release. *Adv. Mater.* **2006**, *18*, 405–409. [CrossRef] [PubMed]

110. Keefer, E.W.; Botterman, B.R.; Romero, M.I.; Rossi, A.F.; Gross, G.W. Carbon nanotube coating improves neuronal recordings. *Nat. Nanotechnol.* **2008**, *3*, 434. [CrossRef] [PubMed]

111. Hirsch, A. Functionalization of single-walled carbon nanotubes. *Angew. Chem. Int. Ed.* **2002**, *41*, 1853–1859. [CrossRef]

112. Shanmugam, S.; Gedanken, A. Electrochemical properties of bamboo-shaped multiwalled carbon nanotubes generated by solid state pyrolysis. *Electrochem. Commun.* **2006**, *8*, 1099–1105. [CrossRef]

113. Wang, K.; Fishman, H.A.; Dai, H.; Harris, J.S. Neural stimulation with a carbon nanotube microelectrode array. *Nano Lett.* **2006**, *6*, 2043–2048. [CrossRef] [PubMed]

114. Edward, D.; Nguyen-Vu, T.B.; Arumugam, P.U.; Chen, H.; Cassell, A.M.; Andrews, R.J.; Yang, C.Y.; Li, J. High efficient electrical stimulation of hippocampal slices with vertically aligned carbon nanofiber microbrush array. *Biomed. Microdevices* **2009**, *11*, 801–808.

115. Su, H.-C.; Lin, C.-M.; Yen, S.-J.; Chen, Y.-C.; Chen, C.-H.; Yeh, S.-R.; Fang, W.; Chen, H.; Yao, D.-J.; Chang, Y.-C. A cone-shaped 3D carbon nanotube probe for neural recording. *Biosens. Bioelectron.* **2010**, *26*, 220–227. [CrossRef] [PubMed]

116. Li, J.; Cassell, A.; Delzeit, L.; Han, J.; Meyyappan, M. Novel three-dimensional electrodes: electrochemical properties of carbon nanotube ensembles. *J. Phys. Chem. B* **2002**, *106*, 9299–9305. [CrossRef]

117. Hsu, H.L.; Teng, I.J.; Chen, Y.C.; Hsu, W.L.; Lee, Y.T.; Yen, S.J.; Su, H.C.; Yeh, S.R.; Chen, H.; Yew, T.R. Flexible UV-ozone-modified carbon nanotube electrodes for neuronal recording. *Adv. Mater.* **2010**, *22*, 2177–2181. [CrossRef] [PubMed]

118. Yen, S.-J.; Hsu, W.-L.; Chen, Y.-C.; Su, H.-C.; Chang, Y.-C.; Chen, H.; Yeh, S.-R.; Yew, T.-R. The enhancement of neural growth by amino-functionalization on carbon nanotubes as a neural electrode. *Biosens. Bioelectron.* **2011**, *26*, 4124–4132. [CrossRef] [PubMed]

119. Li, N.; Zhang, X.; Song, Q.; Su, R.; Zhang, Q.; Kong, T.; Liu, L.; Jin, G.; Tang, M.; Cheng, G. The promotion of neurite sprouting and outgrowth of mouse hippocampal cells in culture by graphene substrates. *Biomaterials* **2011**, *32*, 9374–9382. [CrossRef] [PubMed]

120. Fabbro, A.; Scaini, D.; León, V.n.; Vázquez, E.; Cellot, G.; Privitera, G.; Lombardi, L.; Torrisi, F.; Tomarchio, F.; Bonaccorso, F. Graphene-based interfaces do not alter target nerve cells. *ACS Nano* **2016**, *10*, 615–623. [CrossRef]

121. Zhang, Q.; Xu, J.; Song, Q.; Li, N.; Zhang, Z.; Li, K.; Du, Y.; Wu, L.; Tang, M.; Liu, L. Synthesis of amphiphilic reduced graphene oxide with an enhanced charge injection capacity for electrical stimulation of neural cells. *J. Mater. Chem. B* **2014**, *2*, 4331–4337. [CrossRef]

122. Zhao, S.; Liu, X.; Xu, Z.; Ren, H.; Deng, B.; Tang, M.; Lu, L.; Fu, X.; Peng, H.; Liu, Z. Graphene encapsulated copper microwires as highly MRI compatible neural electrodes. *Nano Lett.* **2016**, *16*, 7731–7738. [CrossRef]

123. Su, H.-C.; Chen, C.-H.; Chen, Y.-C.; Yao, D.-J.; Chen, H.; Chang, Y.-C.; Yew, T.-R. Improving the adhesion of carbon nanotubes to a substrate using microwave treatment. *Carbon* **2010**, *48*, 805–812. [CrossRef]

124. Liu, T.-C.; Chuang, M.-C.; Chu, C.-Y.; Huang, W.-C.; Lai, H.-Y.; Wang, C.-T.; Chu, W.-L.; Chen, S.-Y.; Chen, Y.-Y. Implantable graphene-based neural electrode interfaces for electrophysiology and neurochemistry in in vivo hyperacute stroke model. *ACS Appl. Mater. Interfaces* **2015**, *8*, 187–196. [CrossRef] [PubMed]

125. Collaert, N.; Lopez, C.M.; Cott, D.J.; Cools, J.; Braeken, D.; De Volder, M. In vitro recording of neural activity using carbon nanosheet microelectrodes. *Carbon* **2014**, *67*, 178–184. [CrossRef]

126. Ferguson, J.E.; Boldt, C.; Redish, A.D. Creating low-impedance tetrodes by electroplating with additives. *Sens. Actuators A Phys.* **2009**, *156*, 388–393. [CrossRef] [PubMed]

127. Carretero, N.M.; Lichtenstein, M.P.; Pérez, E.; Cabana, L.; Suñol, C.; Casañ-Pastor, N. IrOx–carbon nanotube hybrids: A nanostructured material for electrodes with increased charge capacity in neural systems. *Acta Biomater.* **2014**, *10*, 4548–4558. [CrossRef]

128. Pérez, E.; Lichtenstein, M.; Suñol, C.; Casañ-Pastor, N. Coatings of nanostructured pristine graphene-IrOx hybrids for neural electrodes: layered stacking and the role of non-oxygenated graphene. *Mater. Sci. Eng. C* **2015**, *55*, 218–226. [CrossRef]

129. Carretero, N.M.; Lichtenstein, M.; Pérez, E.; Sandoval, S.; Tobias, G.; Suñol, C.; Casan-Pastor, N. Enhanced charge capacity in iridium oxide-graphene oxide hybrids. *Electrochim. Acta* **2015**, *157*, 369–377. [CrossRef]

130. Xiao, H.; Zhang, M.; Xiao, Y.; Che, J. A feasible way for the fabrication of single walled carbon nanotube/polypyrrole composite film with controlled pore size for neural interface. *Colloids Surf. B Biointerfaces* **2015**, *126*, 138–145. [CrossRef]

131. Wang, K.; Tang, R.-Y.; Zhao, X.-B.; Li, J.-J.; Lang, Y.-R.; Jiang, X.-X.; Sun, H.-J.; Lin, Q.-X.; Wang, C.-Y. Covalent bonding of YIGSR and RGD to PEDOT/PSS/MWCNT-COOH composite material to improve the neural interface. *Nanoscale* **2015**, *7*, 18677–18685. [CrossRef]

132. Zhou, H.; Cheng, X.; Rao, L.; Li, T.; Duan, Y.Y. Poly (3, 4-ethylenedioxythiophene)/multiwall carbon nanotube composite coatings for improving the stability of microelectrodes in neural prostheses applications. *Acta Biomater.* **2013**, *9*, 6439–6449. [CrossRef]

133. Shi, X.; Xiao, Y.; Xiao, H.; Harris, G.; Wang, T.; Che, J. Topographic guidance based on microgrooved electroactive composite films for neural interface. *Colloids Surf. B Biointerfaces* **2016**, *145*, 768–776. [CrossRef] [PubMed]

134. Kolarcik, C.L.; Catt, K.; Rost, E.; Albrecht, I.N.; Bourbeau, D.; Du, Z.; Kozai, T.D.; Luo, X.; Weber, D.J.; Cui, X.T. Evaluation of poly (3, 4-ethylenedioxythiophene)/carbon nanotube neural electrode coatings for stimulation in the dorsal root ganglion. *J. Neural Eng.* **2014**, *12*, 016008. [CrossRef] [PubMed]

135. Shemesh, O.A.; Tanese, D.; Zampini, V.; Linghu, C.; Piatkevich, K.; Ronzitti, E.; Papagiakoumou, E.; Boyden, E.S.; Emiliani, V. Publisher Correction: Temporally precise single-cell-resolution optogenetics. *Nat. Neurosci.* **2017**, *20*, 1796–1806. [CrossRef] [PubMed]

136. Ran, F.A.; Hsu, P.D.; Wright, J.; Agarwala, V.; Scott, D.A.; Zhang, F. Genome engineering using the CRISPR-Cas9 system. *Nat. Protoc.* **2013**, *8*, 2281. [CrossRef] [PubMed]

137. Ferlauto, L.; Mji, A.L.; Nal, C.; Sca, G.R.; Vagni, P.; Bevilacqua, M.; Wolfensberger, T.J.; Sivula, K.; Ghezzi, D. Design and validation of a foldable and photovoltaic wide-field epiretinal prosthesis. *Nat. Commun.* **2018**, *9*, 992. [CrossRef] [PubMed]

138. Tang, J.; Qin, N.; Chong, Y.; Diao, Y.; Yiliguma; Wang, Z.; Xue, T.; Jiang, M.; Zhang, J.; Zheng, G. Nanowire arrays restore vision in blind mice. *Nat. Commun.* **2018**, *9*, 786. [CrossRef] [PubMed]

 *micromachines*

Article

Constructing a Dual-Function Surface by Microcasting and Nanospraying for Efficient Drag Reduction and Potential Antifouling Capabilities

Liguo Qin [1,2,*], Mahshid Hafezi [1,2], Hao Yang [1,2], Guangneng Dong [1,2] and Yali Zhang [3,*]

[1] Key Laboratory of Education Ministry for Modern Design & Rotary-Bearing System, Xi'an Jiaotong University, Xianning West Road, Xi'an 710049, China

[2] Institute of Design Science and Basic Component, Xi'an Jiaotong University, Xianning West Road, Xi'an 710049, China

[3] Key Laboratory of Biomedical Information Engineering of Ministry of Education, Xi'an Jiaotong University, Xianning West Road, Xi'an 710049, China

* Correspondence: liguoqin@xjtu.edu.cn (L.Q.); yar.lee@xjtu.edu.cn (Y.Z.)

Received: 30 May 2019; Accepted: 21 July 2019; Published: 23 July 2019

Abstract: To improve the drag-reducing and antifouling performance of marine equipment, it is indispensable to learn from structures and materials that are found in nature. This is due to their excellent properties, such as intelligence, microminiaturization, hierarchical assembly, and adaptability. Considerable interest has arisen in fabricating surfaces with various types of biomimetic structures, which exhibit promising and synergistic performances similar to living organisms. In this study, a dual bio-inspired shark-skin and lotus-structure (BSLS) surface was developed for fabrication on commercial polyurethane (PU) polymer. Firstly, the shark-skin pattern was transferred on the PU by microcasting. Secondly, hierarchical micro- and nanostructures were introduced by spraying mesoporous silica nanospheres (MSNs). The dual biomimetic substrates were characterized by scanning electron microscopy, water contact angle characterization, antifouling, self-cleaning, and water flow impacting experiments. The results revealed that the BSLS surface exhibited dual biomimetic features. The micro- and nano-lotus-like structures were localized on a replicated shark dermal denticle. A contact angle of 147° was observed on the dual-treated surface and the contact angle hysteresis was decreased by 20% compared with that of the nontreated surface. Fluid drag was determined with shear stress measurements and a drag reduction of 36.7% was found for the biomimetic surface. With continuous impacting of high-speed water for up to 10 h, the biomimetic surface stayed superhydrophobic. Material properties such as inhibition of protein adsorption, mechanical robustness, and self-cleaning performances were evaluated, and the data indicated these behaviors were significantly improved. The mechanisms of drag reduction and self-cleaning are discussed. Our results indicate that this method is a potential strategy for efficient drag reduction and antifouling capabilities.

Keywords: shark skin; lotus-like structure; drag reduce; antifouling; hydrophobicity

---

1. Introduction

In marine infrastructures, such as ship hulls, wave-energy collectors, and undersea pipelines, high drag forces and biofouling are the two biggest pernicious effects [1–5]. Once fouling settlements are formed on immerged surfaces, they are very difficult to remove, even shortly after their formation. For a forward-moving watercraft, hydrodynamic drag and fuel consumption increase along with the occurrence of biofouling, which significantly decrease the operating speed and cruising distance. Previously, tributyltin self-polishing copolymer coatings were widely used to deal with biofouling on ships [6]. However, they were banned globally for the protection of the ocean environment and marine organisms due to their toxic properties. Thus, developing an effective and environmentally friendly antifouling system has been a subject of immense interest [7]. Many structures, materials, and surfaces observed in nature have inspired researchers to understand their basic principles, such as their intelligence, microminiaturization, hierarchical assembly, and adaptability. Some achievements, for example, rebuilding structures and materials of living creatures, have led to practical applications in the field of materials science and design [8,9].

Shark skin is a well-known example of a material that exhibits efficient drag reduction and antifouling performance. The skin contains minute individual features, called dermal denticles, which are parallel to the swimming direction. It has been proved that these microstructures reduce the formation of vortices, which results in water flowing easily across the shark skin [10]. The water layer near the skin moves faster and reduces the settling time of microorganisms. This eliminates the duration of fouling agglomeration and thus avoids the occurrence of strong adhesion between the substrate and fouling settlements. However, a certain number of microorganisms can still be found on shark skin. Some fouling settlements tend to adhere to particular grooves of shark skin, which is especially serious for artificially replicated shark surfaces. In nature, another well-known example of an antifouling material is the superhydrophobic lotus leaf. Water droplets from rainfall collect contaminated particles and carry them away to protect the leaves from pathogens [11,12]. Wax and hierarchical structures on the lotus leaf are reported to do this. Inspired by lotus leaves, numerous artificial superhydrophobic surfaces have been fabricated on various substrates such as tiles, textile, and glass [13,14]. Further, it was found that lotus-like materials and surfaces also reduce drag. To achieve superhydrophobicity, rough structures, especially the nanomorphologies, play a critical role. Silica is a widely researched nanoparticle which allows the formation of stable nanostructures. By spraying polymethyl methacrylate and a hydrophobic nanoscale silica compound onto a hydrophilic steel substrate, Wang et al. reported that effective drag-reduction behavior was found during a sailing test [15]. Tuo et al. showed that drag reduction was decreased by 20% on a superhydrophobic surface when the flow velocity was between 2 and 5 m/s [16]. However, compared with biomimetic shark-skin surfaces, the drag reduction of artificial lotus-like surfaces was much lower, and the mechanism of this effect is still under investigation.

Recently, different technologies that introduce bio-inspired functions have been investigated to create surfaces that exhibit excellent and synergistic performances similar to living organisms. Additive manufacturing was used to design microriblet features, and a 3D-printed shark skin demonstrated almost 10% average fluid drag reduction [17]. Lu used chemical deposition to fabricate a micro–nanohierarchical structure for superhydrophobicity and drag reduction. The results indicated that the nanostructure may help the superhydrophobic surface exhibit drag reduction properties [18]. However, the connection between microscale and nanoscale features (such as the surface functions of shark skin and lotus-like structures) has not been completely revealed [12,19,20]. It is believed that superimposing patterns of a shark-skin surface (BS) and a lotus-like structure (LS) at the microscale and nanoscale, respectively, would produce a dual-function surface possessing both efficient drag reduction and antibiofouling performance. Therefore, the aim of this study was to develop a bio-inspired shark-skin and lotus-structure (BSLS) surface and for use in marine equipment applications. To achieve this, the hierarchical structure was introduced on polyurethane (PU) polymer. A vacuum casting procedure was selected to produce shark-skin-like micromorphologies. Mesoporous silica particles of a size similar to lotus-like structures were deposited on the replicated shark surface throughout the spraying process. The biomimetic surfaces were evaluated with various measurements. The morphologies were observed by scanning electron microscopy. The surface wettability was conducted by water contact angle (CA) measurements. Finally, antifouling, self-cleaning, and liquid drop impact experiments were carried out to illustrate the mechanism of their synergistic effect.

2. Materials and Methods

The tiger shark (Heterodontus japonicus) was chosen for the template and was obtained from a fisherman in Xiamen, China. Shark skin from the posterior part was selected. 3-(aminopropyl) triethoxysilane (APTES) was used to disperse the nanoparticles and was purchased from Shanghai Macklin Biochemical Co., Ltd (Shanghai, China). Liquid polyurethane (PU, Sigma-Aldrich, Shanghai, China) was selected to replicate the shark skin. It was constituted with a two-component precursor and curing agent. Mesoporous silica nanospheres (MSN, diameter = 185 ± 30 nm) were purchased from Rizhao biotech company. The glutaraldehyde, alcohol, and acetone reagents were all analytical grade.

BS samples were fabricated by a typical vacuum casting process. Prior to the casting process, pretreatment of the shark skin was conducted. The soft tissue under the rigid shark skin was carefully removed. After the skin template was flattened and fixed to a rigid mold, it was immersed in 5% glutaraldehyde for over 2 h. Then, the template was rinsed with deionized water at least three times. Graded ethanol was used to conduct the dehydration process, which avoided deformation from excessive water loss during the typical drying process. Repeated rinsing was performed and concentrations of 30%, 50%, 75%, 80%, 95%, and 100% ethanol solution were applied every 20 min for each step. After keeping the template in a thermostat at 48 °C (Kewei 101, Beijing, China) for over 12 h, the pretreated template was ready for replication.

The detailed procedure for fabricating the BSLS surface is illustrated in Figure 1. Firstly, the treated template was localized on a plastic plate. A compound of liquid silicon precursor and curing agent was poured and the average of thickness of the liquid silicon was controlled to 2.5 mm. In order to make sure the liquid silica was fully adapted to the shark-skin template, all processes were conducted in a vacuum chamber. By carefully stripping the template, the silicon negative mold was obtained. Liquid polyurethane was poured and cured to complete the precise positive duplication. The third step was to spray the modified silica particles onto the replicated shark-skin surfaces to introduce the lotus structure. For the modification, mesoporous silica nanospheres were first dispersed uniformly in acetone solution. APTES with a concentration of 5% was added and used to modify the MSN. After sonication for 5 min, a good dispersion of MSNs was achieved and the radical groups of APTES strongly bonded the nanoparticles to the substrates [21]. The dispersed nanoparticles were sprayed on the BS surface by a spraying gun. Different concentrations were used to achieve various coverages of 3.75, 4.25, 5, and 7.5×10^{-3} mg/mm^2. Samples were denoted as BSLS1–BSLS4, respectively.

The morphologies of the real shark skin, the negative mold, and the replicated PU samples were observed by field emission scanning electron microscopy (SEM, MAIA3 LMH, TESCAN, Warrendale, PA, USA) and 3D laser confocal microscopy (OLS4000, Olympus, Tokyo, Japan). The well-dispersed mesoporous silica nanospheres were checked by transmission electron microscopy (TEM, JEOL JEM-2100Plus, Tokyo, Japan). The chemical composition of the silica nanospheres was obtained with an energy-dispersive spectrometer (EDS, OXFORD instruments, Abingdon, UK) and attenuated total reflectance–Fourier transform infrared spectroscopy (ATR-FTIR) (TENSOR27, Bruker, Karlsruhe, Germany).

Static CA was measured by a video-based contact angle system (JC2000D2A, Powereach, Shanghai, China) at room temperature. Young–Laplace fitting was applied to calculate the CA value. Three different regions were tested, and mean CA was regarded as the apparent CA. The effect of the proposed BSLS approach on the dynamic surface wettability was analyzed. The advancing contact angle and receding contact angle were calculated to obtain the contact angle hysteresis (CAH). For each CAH measurement, three different positions for each sample were tested.

The drag reduction of the BSLS surface was evaluated by a rheometer apparatus as described in [22,23]. Figure 2 shows the overview of the rheometer apparatus (MCR302, Anton Paar, Graz, Austria). The sample was fixed underneath the rotary plate. To keep the denticles in the same direction relative to the flow, the samples was cut into small parts and then they were oriented in the circle direction in the rheological experiment. We then carefully used the glue to adhere each part. Artificial seawater (1 mL) was injected into the gap of the rotary plate and BSLS surface. Here, the gap (h) was set to 1 mm and the rotation direction was along with the shark denticles. The velocity of the upper rotor plate was set to 100 rpm, and the drag reduction ratio was measured using the following equation:

$$\text{Drag reduce } \% = \frac{\tau_{\text{nonslip}} - \tau_{\text{slip}}}{\tau_{\text{nonslip}}}$$

where $\tau_{nonslip}$ and τ_{slip} represent the shear stresses at the wall when no-slip and slip boundary conditions were used, respectively.

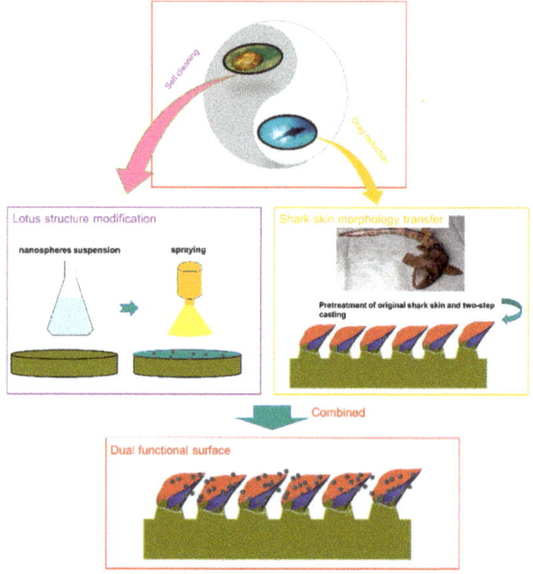

Figure 1. Process of synthetic replication of shark skin and lotus surface.

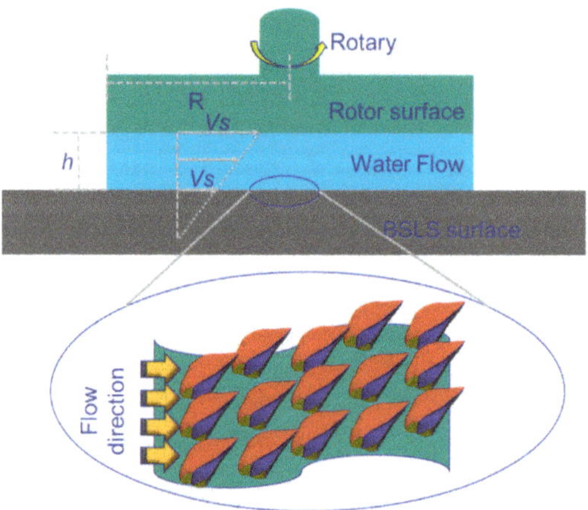

Figure 2. The schematic of the rheometer apparatus for measuring drag reduction, for which dermal denticles were oriented along the water flow.

Self-cleaning properties were investigated by contaminating the BSLS surface with simulated pollution. Hydrophilic silicon carbide (SiC, Sigma-Aldrich, Shanghai, China) particles were used as simulated pollution. The size of the SiC particles ranged from 10 to 15 µm. SiC particles were chosen because of their performance similarity to natural dirt (shape, size, and hydrophilicity). SiC particles (300 mg) were gently sprayed onto the samples. Samples were localized on a tilted plate (30°) and water droplets were dripped from a specified height (30 ± 2 mm). The dripping was controlled at a constant rate (total duration of 1 min using 5 mL of water). The self-cleaning efficiency was calculated according to the weight variation before and after dripping.

Antifouling properties were evaluated by protein adsorption and protozoan colonization. For protein adsorption, samples were cut into blocks with size of 72 mm^2 (6 by 12 mm) and soaked in a Bovine Serum Albumin (BSA) solution (0.5 mg/mL and pH ~7.4) for 6 h at room temperature. Using the Bradford method [24], the residual of BSA within the solution was measured. By subtracting its initial concentration, the adsorption of BSA was calculated. An ultraviolet spectrophotometer (UV-3600, Shimadzu Co., Kyoto, Japan) with a wavelength of 595 nm was used to measure the absorbance of the solutions. The amount of adsorbed BSA protein on the samples was represented in the form of µg/cm^2. For protozoan colonization, samples were cocultured with protozoan suspension with a concentration of 1×10^3 protozoans/mL. After 48 h, samples were taken out, washed with distilled water, and observed under an optical microscope. The number of viable protozoans was counted and three different points were measured for each sample.

To evaluate the robustness of the biomimetic surface, samples were exposed to water with continuous impacting. The velocity was set to 1.0 m/s. After every hour, samples were taken out and dried totally. The contact angle was measured. For all the wettability tests, 5 µL of distilled water was dipped on the specimens by a microliter syringe, and each specimen was tested thrice to obtain the mean value. We further measured the mechanical robustness of the BSLS surface. Sandpaper abrasion was used to study the wear resistance properties. A 1 by 1 cm^2 sample was fixed on the bottom of a 50 g weight (4.9 KPa). The sample was then uniformly slid onto 600 grit sandpaper (Zhangshi Co., Shenzhen, China). Each abrasion cycle spanned 10 cm in a forward motion. After 100 cycles, the worn morphologies were observed under SEM.

3. Results and Discussion

3.1. Surface Characterization

Mesoporous silica nanospheres were selected due to their morphological similarity to lotus-like structures and ease of chemical modification with biomolecules. Additionally, mesoporous silica nanospheres are theranostic agents and carriers for drug delivery [25,26], which could be considered as a potential application in loading antifouling agents for release. TEM was used to observe the fine features of the mesoporous silica nanospheres. As shown in Figure 3a,c, the diameter of the nanospheres was in the range of 162–185 nm and a hollow structure was observed. After modification with APTES, the diameter was in range of 228–261 nm (Figure 3b,d). The EDS map of one single nanosphere confirmed that it mainly consisted of Si and O elements (Figure 3e). Carbon elements were also observed, which indicated that APTES was successfully grafted on the silica nanosphere. The FTIR spectra of modified silica nanospheres are shown in Figure 3f. The typical absorption peaks of SiO$_2$ materials were observed, such as Si–O–Si symmetric stretching (798 cm^{-1}), Si–OH asymmetric vibration (950 cm^{-1}), and Si–O asymmetric vibration (1108 cm^{-1}) [27]. The peak at 2950 cm^{-1} was the C–H stretching, which further confirmed the grafting of APTES.

Figure 4 shows the morphologies of the original shark skin, the negative mold, and the replicated PU substrate. The replicated PU sample (Figure 4c) exhibited almost the same surface features as that of the denticles on the real shark. Figure 4b shows the negative mold of silicone, which showed more well-defined structures and higher integrity than the original shark skin. This was attributed to the good seepage characteristic of liquid silicone, which was capable of filling in the cavities of the mold before it was fully cured. As shown in Figure 4f, the horizontal curve section describes the central section of one single denticle (red line marked in Figure 4e). The pitch of the groove was around 70 µm and the height was around 25 µm. The results indicated the high replication precision of geometrical morphology, and the obliquity of the scales was maintained during the fabrication process.

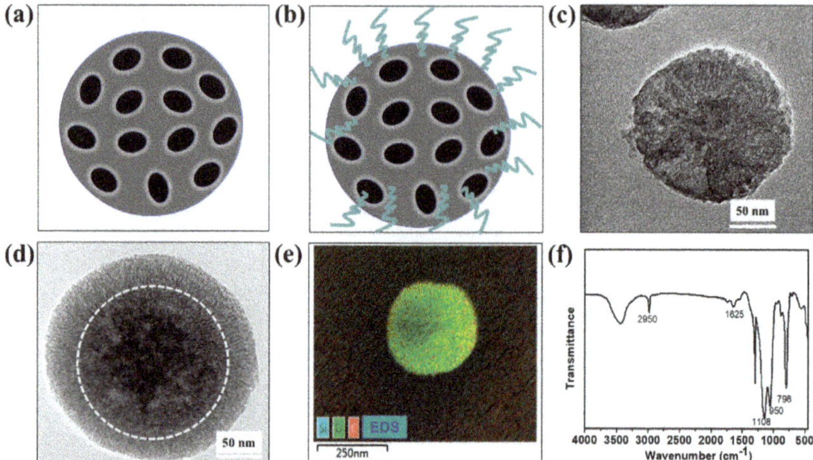

Figure 3. Mesoporous silica nanospheres used in the study. (**a**,**b**) are illustrations of bare mesoporous silica nanospheres and mesoporous silica nanospheres modified with 3-(aminopropyl) triethoxysilane (APTES). (**c**,**d**) are the corresponding transmission electron microscopy (TEM) images, (**e**) chemical composite, and (**f**) Fourier transform infrared spectroscopy (FTIR) spectrum of mesoporous silica nanospheres modified with APTES.

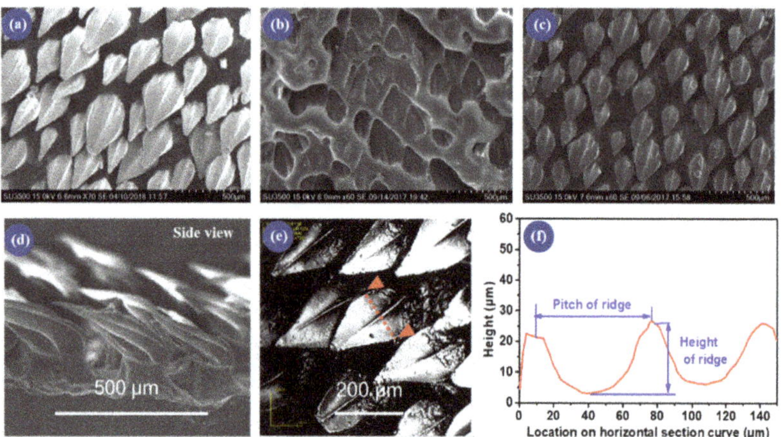

Figure 4. SEM and surface profile of shark skin, negative mold, and replicated polyurethane (PU) coating: (**a**) pretreated shark skin, (**b**) negative mold, (**c**) replicated shark surface, (**d**) side view of dermal denticles, (**e**) confocal microscope image, and (**f**) the section profile of dermal denticles that was marked in (**e**).

Micrographs of nanospheres at different concentrations on dermal denticles are shown in Figure 5. A homogeneous distribution of mesoporous silica nanospheres was observed on all samples. As the spraying concentration of mesoporous silica nanospheres increased, more lotus-like structures were found on dermal denticles. The results imply that the BSLS surface possessing dual biomimetic morphologies was successfully fabricated. By combining microcasting and spraying, the BSLS surface exhibited the surface features of shark skin and hierarchical lotus-like structures. By adjusting the concentration of MSN, various coverages of lotus-like structures were achieved.

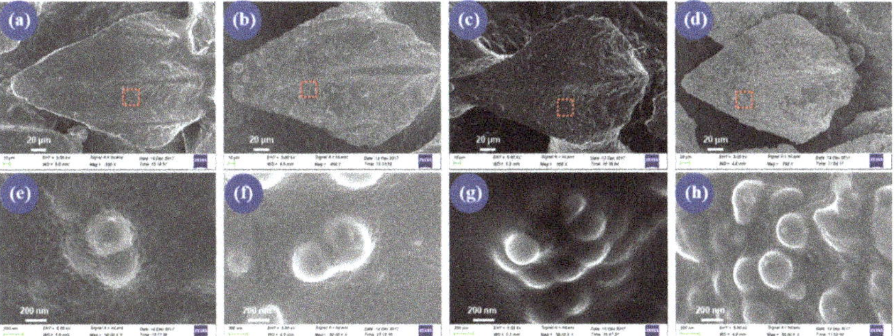

Figure 5. SEM micrographs were taken at a 10° tilt angle showing the morphologies of the shark-skin and lotus-structure (BSLS) samples with different concentration modifications of mesoporous silica nanospheres: (**a–d**) represent 3.75, 4.25, 5, and 7.5 × 10^{-3} mg/mm^2, respectively; (**e–h**) correspond to the red area in (a–d).

3.2. Surface Wettability

To investigate the effect of mesoporous silica nanospheres on the surface wettability, the static contact angle, and contact angle hysteresis were measured on the BSLS samples. As shown in Figure 6a, flat PU had a contact angle around 83.1°, indicating hydrophilicity. The contact angle of the replicated shark-skin surface was 118.1° and increased by 42% compared with that of flat PU. The contact angle hysteresis for flat PU and replicated shark skin was higher than 30°, which indicated that there was high adhesion between the surface and water. After being sprayed with nanospheres, the contact angles for all substrates were increased. When the flat surface was modified with MSNs, the surface wettability improved slightly, for example, 86.6° for LS1 and 98.2° for LS2. For the replicated shark skin, the contact angle increased dramatically to 147.2° with the MSN modification (e.g., the BSLS2 sample) and the surface was close to being superhydrophobic. Importantly, this combination has been shown to create a rough surface structure and low-surface-energy nanospheres when fabricating superhydrophobic surfaces. As it was observed, the dual-structured surface for the contact angle hysteresis had a lower value compared with that of the flat and replicated shark-skin surfaces, respectively.

To effectively fabricate a water-repelling surface, an effective air film between the surface and water droplets is fundamental. Here, air was easily trapped by the rough structure of the samples. The wetting mode of the hydrophilic surface could be considered using the Cassie–Baxter wetting mode, which is mainly used to describe a heterogeneous wetting regime [28]. In this study, the prepared BSLS surface was considered as a typical composite solid–air surface. Proof was provided by the Cassie–Baxter equation [29,30]:

$$\cos\theta_w = \Phi_s(\cos\theta_e + 1) - 1$$

where Φ_s is the area ratio between liquid and solid contact. It built the relationship between the apparent contact angle (θ_w) of the composite interface and its intrinsic contact angle (θ_e). The surface became more hydrophobic as Φ_s increased. A large fraction of air could be trapped by interstices of the microstructures. Water droplets were mainly localized on peaks of the surface and had difficulty penetrating the interstices between the different topographical peaks. With the MSN modification, the synergistic effect of microroughness and MSNs may have caused the lower CAH. These results indicate that surface morphology, especially the nanoscale features, had a more significant effect on controlling the surface wettability. However, this enhancement was not obvious when the number of nanospheres was higher than 5 × 10^{-3} mg/mm^2. This could have been due to fewer morphological changes if the concentration of nanospheres was increased to a certain value and a certain amount of aggregation. At the same time, the structures, such as those shown in Figure 5c,d, did not trap sufficient air in the structure for a low CAH and high CA.

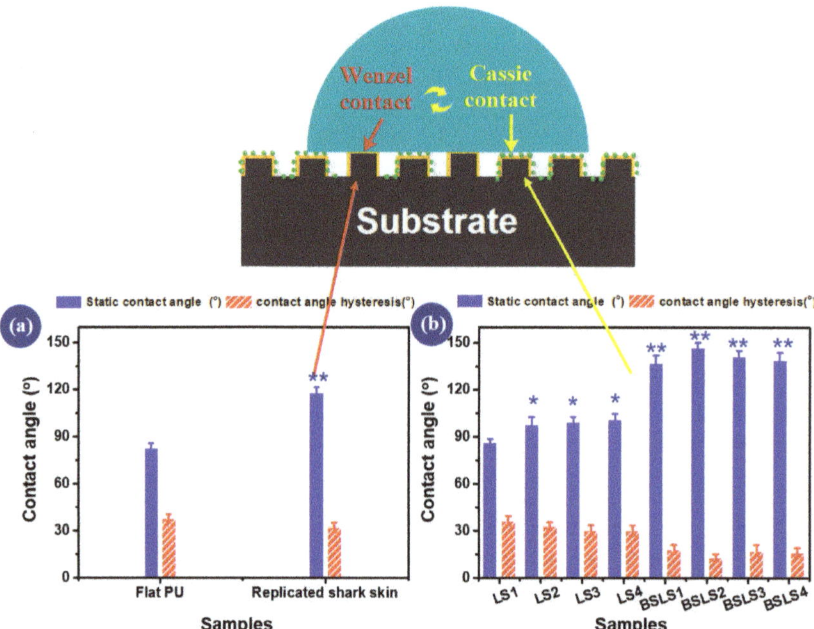

Figure 6. Bar charts showing the measured static contact angle and contact angle hysteresis. Four different spraying concentrations were used for the flat PU and replicated shark skin: 3.75, 4.25, 5, and 7.5×10^{-3} mg/mm^2. Samples were denoted as lotus-like structure LS1–LS4 and BSLS1–BSLS4, respectively. (**a**) static contact angle and contact angle hysteresis for Flat PU and BS samples, (**b**) static contact angle and contact angle hysteresis after spraying SiO$_2$ * $p < 0.05$ different and ** $p < 0.01$ significantly different from contact angle (CA) measured on flat PU by One-way ANOVA.

3.3. Drag Reduction Performance

The drag reduction behavior of the BSLS surface was characterized by a rheometer apparatus. Figure 7 shows the shear stress under constant flow velocity in the simulated seawater environment. In the test condition, the mean shear stress of flat PU was 5.41 Pa, which was the highest among all samples. For the replicated shark surface, the mean shear stress was 3.73 Pa. It decreased by 31%, exhibiting a good reduction of shear stress and drag force. When the replicated shark surface was further modified with MSNs, all surfaces showed lower shear stress values compared with that of flat PU. For a lower concentration modification, the shear stress was decreased (i.e., BSLS1 and BSLS2) even less than that of the replicated shark surface. The drag reduction of BSLS2 was lower than 36.7%, which was near that of the original shark skin (39.0%). When the concentration was further increased (i.e., BSLS3 and BSLS4), the shear stress decreased less. The drag reduction was 20.2% and 18.5% for BSLS3 and BSLS4, respectively. Generally, the relative velocity was considered to be zero at the boundary between the solid wall and liquid. This was widely characterized as a no-slip boundary condition. In our study, the velocity of fluid appeared to a nonzero for BSLS surfaces. An air layer was considered to form in interstices of the biomimetic surface. This layer acted as an air pad and a shear-reducing boundary was formed between the solid surface and the fluid, which significantly reduced the drag force. Previous works have shown that a shark-skin-like surface encourages the formation of secondary small eddies inside the riblet spacing [31–33]. A more detailed investigation of the effect of the BSLS surfaces on drag reduction is presently being carried out.

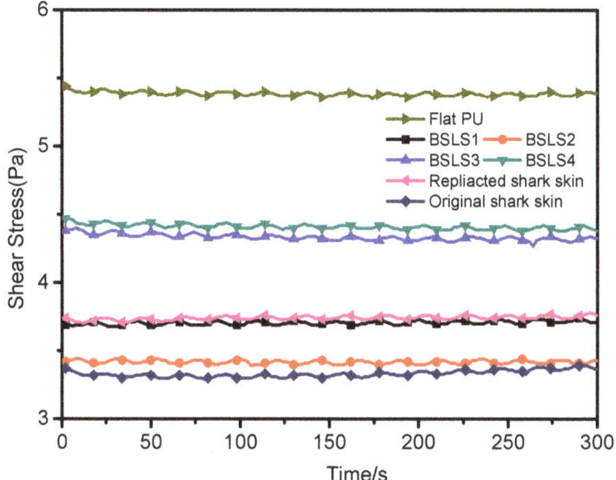

Figure 7. The shear stress for different samples.

3.4. Antifouling and Self-Cleaning Performance

Protein adsorption is the first step of biofouling, which is undesirable for most marine applications. Evaluating the adsorption amount of protein on a surface is regarded as a common way of estimating antibiofouling performance. For the antifouling study, the BSA adsorption of flat PU, replicated shark skin, SiO_2-nanosphere-modified PU, and BSLS samples was measured. As shown in Figure 8, the BSA adsorption varied for the different samples. For the replicated shark-skin surface, the microroughness increased the BSA adsorption (44.2 $\mu g/cm^2$) due to the enhanced surface contact area. The microroughness provided more sites for the adsorption of protein molecules [22]. With SiO_2 nanosphere modification, the amount of BSA adsorption decreased. The lowest adsorption level was observed on BSLS2 (8.75 versus 39.27 $\mu g/cm^2$ for flat PU), indicating a significant enhancement in resistance to protein adhesion. This can be attributed to the enhanced hydrophobicity due to the combination of replicated shark skin and SiO_2 nanospheres. As a result of nanoscale roughness (SiO_2 nanospheres), protein resistance was expected to be increased due to the water barrier, which would be caused by the enhancement of superhydrophobicity [34,35]. A physical barrier would form to prevent direct interaction between the surface and the protein. The intermediate wettability of the nanostructured surfaces would promote and induce the conditions for forming protein aggregates and nucleation. In this study, the increase in nanoscale roughness alongside the enhancement of hydrophilicity was the main reason (e.g., decreasing ~78% of protein adsorption for biomimetic surfaces (BSLS2)). Additionally, the surface negative charge was improved due to the hydrolyzed hydroxyl groups, which were derived from the deposited SiO_2 nanoparticles. These generated a synergistic improvement of protein resistance [36,37]. Meanwhile, the sliding time of droplet was measured when samples was fixed on the tiled platform. As shown in Movie S1 (Supplementary Materials. Movie S1), for the flat and LS samples, both of water and milk were stuck on the surface. On the contrary, less adhesion and short sliding time were observed on the replicated shark skin and BSLS surface. On a 12 mm sliding distance, it took 610 ms for the replicated shark skin while it took 360 ms for BSLS surface when the test medium was water. Similar phenomenon was observed when the test medium was milk.

Settled protozoans were visualized by a Zeiss optical microscope. As shown in Figure S1 (Supplementary Materials. Figure S1), the highest density of settled protozoans was observed on the BS surface. A significantly lower density of protozoans was found on the LS surface. Compared with the flat sample, the settlement density was further reduced on the BSLS surface. Protozoa adhesion was a dynamic process and was influenced by various factors. The microstructure of BS provided a sufficient location for protozoa settlement. Protozoans accumulated on the BS surface. After SiO_2 was deposited, the physical barrier and lower surface energy made the BSLS surface ideal for antifouling.

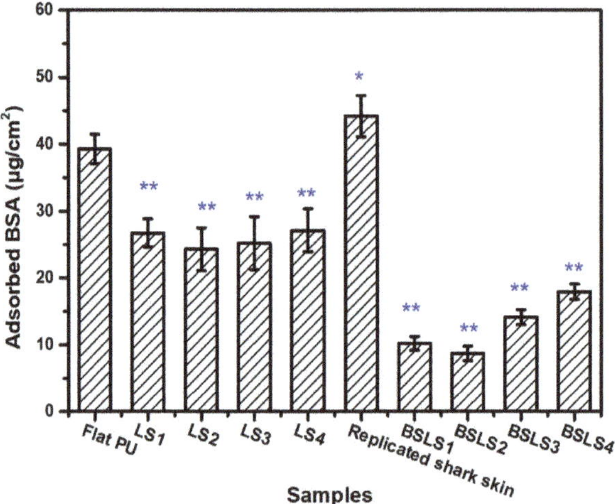

Figure 8. The amount of adsorbed BSA on the surface measured by the Braford method. * $p < 0.05$ different and ** $p < 0.01$ significantly different from adsorbed BSA measured on flat PU by One-way ANOVA.

The self-cleaning performance of the samples was evaluated by measuring the residual mass of the pollutants after washing. The reduced residual amount was due to the better antifouling performance of the samples. As shown in Table 1, the residual mass of pollutants for the modified surface decreased compared with that of the flat surface. The residual mass of pollutants was 13.4 mg for the flat surface. Due to MSN spraying, the residual mass of the contaminant was significantly decreased, almost less than 30% of the flat sample. For the BS surface, the residual mass was slightly reduced by about 23%. However, for the dual biomimetic BSLS surface (e.g., BSLS2), the pollutants were almost removed.

Table 1. Measured residual mass of the samples.

Samples	Flat PU	LS1	LS2	LS3	LS4	Replicated Shark Skin	BSLS1	BSLS2	BSLS3	BSLS4
Δm (mg)	13.4 ± 2.3	12.1 ± 2.4	12.5 ± 2.6	13.2 ± 3.1	13.1 ± 2.9	9.2 ± 1.9	3.5 ± 0.9	3.1 ± 0.8	3.4 ± 1.1	4.1 ± 1.2

3.5. Surface Duration

Considering most practical applications, functionality and superhydrophobicity must remain stable for a long time, which is very difficult to achieve. To evaluate their robustness, the samples were exposed to a sustained water flow. The speed was controlled at 1.0 m/s and the duration of impacting was 10 h. The evolution of the contact angle is shown in Figure 9a. It can be observed that the surface wettability of BSLS was quite stable. After 10 h of water flow, the contact angle slightly decreased from 156° to 145°. Commonly, the hydrophobic property is gradually weakened after exposure to humid conditions for a few hours. Also, if MSNs were lost or removed by water impacting, it would influence the performance of the BSLS surface (e.g., a sharp decrease of the water contact angle). The results showed that the finished surface was almost the same hydrophobicity of the initial surface. There was little damage to the nanomorphologies after a long duration of flow impacting, which indicated that the BSLS surface exhibited long-term stabilization. The integrity of the exposed surface was further verified by SEM, as shown in Figure 9b,c. The top and section views confirmed that the biomimetic surface retained a similar profile to when it was prepared. One reason for this was the strong interaction between SiO_2 nanospheres and the replicated shark skin. Another reason was the relevant reduction of flow drag caused by the effect of the dermal denticles. An interesting observation was that when the water jet impacted the samples, it rapidly bounced off their surface. This phenomenon indicated that the typical Cassie–Baxter wetting state existed on the biomimetic surfaces, which extremely weakened the interaction between the BSLS surface and water. Thus, less damage occurred.

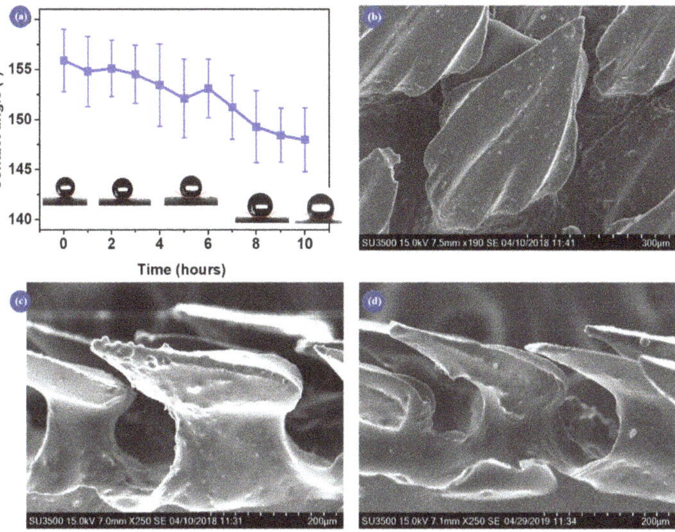

Figure 9. Durability of the biomimetic surface. (**a**) The evolution of contact angle with water impacting, (**b,c**) top and section views of the surface after 4 hours' impacting, and (**d**) section view of the surface after 10 hours' impacting. The sample of BSLS2 was used.

We further measured the mechanical robustness of the BSLS surface. As shown in Figure 10, the flat PU was covered with various ploughs. With MSN modification, the wear resistance was significantly improved and less ploughs were observed. For the BS surface, several riblets were severely damaged, while the surface remained stable for the BSLS sample (Figure 10d). Compared with the flat surface, the wear resistance of the BSLS surface was greatly improved.

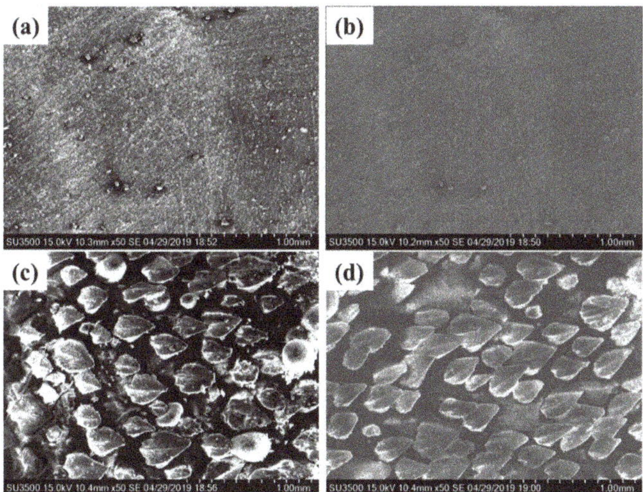

Figure 10. SEM of surfaces after abrasion with sand paper: (**a**) flat sample, (**b**) LS sample, (**c**) BS sample, and (**d**) BSLS sample.

3.6. Antidrag and Antifouling Mechanism

The experimental evidence showed that the BSLS surface with an optimized nanosphere concentration achieved significant drag-reduction and fouling-resistance effects, thus demonstrating its potential application in decreasing energy consumption during sailing. Many studies have confirmed that surface wettability is regulated by surface roughness [30,38]. In this study, the biomimetic morphology was fabricated on PU. Superhydrophobicity was achieved by the synergistic effect of the replicated shark skin and MSNs. Previous researchers have revealed that drag is mainly influenced by the slippage between the liquid and the contacted solid [23,39]. For the hydrophobic surface, air layers existed according to the Cassie model. When the biomimetic sample was immersed in distilled water, the surface was wrapped by different types of bubbles, as shown in Figure 11. As marked in the red area, small bubbles were mainly observed on the LS surface, while big bubbles were mainly observed on the replicated shark-skin surface. For the BSLS samples, both small and big bubbles were observed. The slippage of the specimen underwater was the result of interactions among the molecules of the solid surface, air bubbles, and liquid, which were regulated by the surface roughness, surface wettability, and shearing rate. Commonly, air bubbles between the solid and liquid interface is the main reason for increasing the slip length.

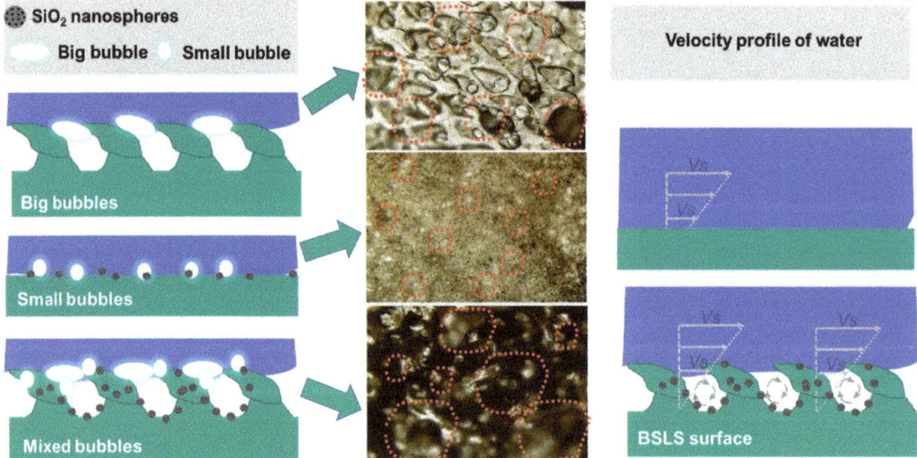

Figure 11. Antidrag and antifouling mechanism.

Combining the results of the water impacting experiments and the contact angle measurements, the slipping of water indicated that the chance of forming large air bubbles was improved on the replicated surface, especially for the BSLS surface. This enhanced effect makes the liquid and even contaminants more easily removed from the top of the contacted surfaces. As described in our and others' research, kinematic viscosity notably declined and, thus, self-cleaning was achieved, which converted the solid–liquid contact mode into the solid–air–liquid mode [40]. Moreover, small bubbles merge in bigger bubbles or thick air layers, which acted as a barrier or air pillow and caused adhesion to lessen [26,41].

4. Conclusions

To create durable bio-inspired surfaces with superhydrophobic, antidrag, and self-cleaning properties, dual biomimetic morphologies were developed for use on commercial PU. Two steps were involved in the fabricating process: Microcasting and nanosphere spraying, which created both the shark-skin microdenticles and the lotus-like nanofeatures. The results showed that it was possible to realize a larger area and high replication precision by microcasting. The subsequent spraying made the samples exhibit hybrid micro/nanostructures with excellent replication of the shark-skin microstructure. The spraying amount was the key factor affecting and controlling the CA and CAHs. At the concentration of 4×10^{-3} mg/mm^2, the dual biomimetic surface showed near superhydrophobicity and the lowest contact angle hysteresis. The water contact angle was 147°, which was 29° higher than that of the unsprayed BS surface. The BSLS2 specimen exhibited excellent drag-reduction and self-cleaning performances. Compared with flat PU, the drag reduction and self-cleaning for BSLS2 were almost increased by 36.7% and 76%, respectively. The robustness of the BSLS surface was verified by sandpaper abrasion and long-term water impacting. The air-bubble layer that existed at the contact surface was believed to have had a significant impact on the drag-reduction and self-cleaning performances. This surface manufacturing technique is fast and feasible and results in the surface possessing better self-cleaning, antifouling, and drag-reduction capabilities. It is believed that this method is an effective strategy that has promising industrial applications for practical utilizations.

Supplementary Materials: The following are available online at http://www.mdpi.com/2072-666X/10/7/490/s1, Figure S1: Optical images of protozoa adhesion on different samples. (a) flat sample, (b) LS, (c) BS, and (d) BSLS. The scale bar is 300 μm. The red circle shows the location of protozoa. Video S1: Sliding tests of the dual bio-inspired shark-skin and lotus-structure surfaces.

Author Contributions: Conception, L.Q., G.D., and Y.Z.; Performed the experiments, H.Y.; Data analysis, M.H.; Wrote the paper, L.Q.

Funding: This work was supported by the Fundamental Research Funds for the Central Universities. The authors acknowledge the joint financial support from the National Key R&D Program of China (2018YFB1306100), National Natural Science Foundation of China (51605370), Postdoctoral Science Foundation of Shaanxi Province (2017BSHEDZZ122), China Postdoctoral Science Foundation funded project (2016M602802), and the Natural Science Fund of Shaanxi Province (2017JQ5009).

Acknowledgments: We thank Zijun Ren at the instrument analysis center of Xi'an Jiaotong University for his assistance with SEM and TEM analysis.

Conflicts of Interest: The authors declare no conflict of interests.

References

1. Liu, X.; Wu, W.; Wang, X.; Luo, Z.; Liang, Y.; Zhou, F. A replication strategy for complex micro/nanostructures with superhydrophobicity and superoleophobicity and high contrast adhesion. *Soft Matter* **2009**, *5*, 3097–3105. [CrossRef]

2. He, X.; Cao, P.; Tian, F.; Bai, X.; Yuan, C. Autoclaving-Induced in-Situ Grown Hierarchical Structures for Construction of Superhydrophobic Surfaces: A New Route to Fabricate Antifouling Coatings. *Surf. Coat. Technol.* **2019**, *357*, 180–188. [CrossRef]

3. Hall-Stoodley, L.; Costerton, J.W.; Stoodley, P. Bacterial biofilms: From the Natural environment to infectious diseases. *Nat. Rev. Microbiol.* **2004**, *2*, 95–108. [CrossRef] [PubMed]

4. Singh, R.; Nadhe, S.; Wadhwani, S.; Shedbalkar, U.; Chopade, B.A. Nanoparticles for Control of Biofilms of Acinetobacter Species. *Materials* **2016**, *9*, 383. [CrossRef] [PubMed]

5. Ji, Y.; Sun, Y.; Lang, Y.; Wang, L.; Liu, B.; Zhang, Z. Effect of CNT/PDMS Nanocomposites on the Dynamics of Pioneer Bacterial Communities in the Natural Biofilms of Seawater. *Materials* **2018**, *11*, 902. [CrossRef] [PubMed]

6. Yebra, D.M.; Kiil, S.; Dam-Johansen, K. Antifouling technology—Past, present and future steps towards efficient and environmentally friendly antifouling coatings. *Prog. Org. Coat.* **2004**, *50*, 75–104. [CrossRef]

7. Kumar, A.; Mills, S.; Bazaka, K.; Bajema, N.; Atkinson, I.; Jacob, M.V. Biodegradable optically transparent terpinen-4-ol thin films for marine antifouling applications. *Surf. Coat. Technol.* **2018**, *349*, 426–433. [CrossRef]

8. Mitragotri, S.; Lahann, J. Physical approaches to biomaterial design. *Nat. Mater.* **2009**, *8*, 15–23. [CrossRef]

9. Barthlott, W.; Mail, M.; Neinhuis, C. Superhydrophobic hierarchically structured surfaces in biology: Evolution, structural principles and biomimetic applications. *Philos. Trans. R. Soc. A Math. Phys. Eng. Sci.* **2016**, *374*, 20160191. [CrossRef]

10. Liu, Y.H.; Li, G.J. A New Method for Producing Lotus Effect on a Biomimetic Shark Skin. *J. Colloid Interface Sci.* **2012**, *388*, 235–242. [CrossRef]

11. Liu, K.; Jiang, L. Bio-inspired design of multiscale structures for function integration. *Nano Today* **2011**, *6*, 155–175. [CrossRef]

12. Koch, K.; Bhushan, B.; Barthlott, W. Diversity of structure, morphology and wetting of plant surfaces. *Soft Matter* **2008**, *4*, 1943. [CrossRef]

13. Bixler, G.D.; Theiss, A.; Bhushan, B.; Lee, S.C. Anti-fouling properties of microstructured surfaces bio-inspired by rice leaves and butterfly wings. *J. Colloid Interface Sci.* **2014**, *419*, 114–133. [CrossRef]

14. Liu, K.; Jiang, L. Bio-Inspired Self-Cleaning Surfaces. *Annu. Rev. Mater. Res.* **2012**, *42*, 231–263. [CrossRef]

15. Wang, N.; Tang, L.; Cai, Y.; Tong, W.; Xiong, D. Scalable superhydrophobic coating with controllable wettability and investigations of its drag reduction. *Colloids Surf. A Physicochem. Eng. Asp.* **2018**, *555*, 290–295. [CrossRef]

16. Tuo, Y.; Chen, W.; Zhang, H.; Li, P.; Liu, X. One-Step Hydrothermal Method to Fabricate Drag Reduction Superhydrophobic Surface on Aluminum Foil. *Appl. Surf. Sci.* **2018**, *446*, 230–235. [CrossRef]

17. Li, Y.; Mao, H.; Hu, P.; Hermes, M.; Lim, H.; Yoon, J.; Luhar, M.; Chen, Y.; Wu, W. Bioinspired Functional Surfaces Enabled by Multiscale Stereolithography. *Adv. Mater. Technol.* **2019**, *4*, 1800638. [CrossRef]

18. Lu, Y. Fabrication of a lotus leaf-like hierarchical structure to induce an air lubricant for drag reduction. *Surf. Coat. Technol.* **2017**, *331*, 48–56. [CrossRef]

19. Bixler, G.D.; Bhushan, B. Bioinspired micro/nanostructured surfaces for oil drag reduction in closed channel flow. *Soft Matter* **2013**, *9*, 1620–1635. [CrossRef]

20. Jaggessar, A.; Shahali, H.; Mathew, A.; Yarlagadda, P.K.D.V. Bio-mimicking nano and micro-structured surface fabrication for antibacterial properties in medical implants. *J. Nanobiotechnol.* **2017**, *15*, 64. [CrossRef]

21. West, J.; Critchlow, G.; Lake, D.; Banks, R. Development of a superhydrophobic polyurethane-based coating from a two-step plasma-fluoroalkyl silane treatment. *Int. J. Adhes. Adhes.* **2016**, *68*, 195–204. [CrossRef]

22. Wu, Y.; Cai, M.; Li, Z.; Song, X.; Wang, H.; Pei, X.; Zhou, F. Slip flow of diverse liquids on robust superomniphobic surfaces. *J. Colloid Interface Sci.* **2014**, *414*, 9–13. [CrossRef] [PubMed]

23. Wu, Y.; Xue, Y.; Pei, X.; Cai, M.; Duan, H.; Huck, W.T.S.; Zhou, F.; Xue, Q. Adhesion-Regulated Switchable Fluid Slippage on Superhydrophobic Surfaces. *J. Phys. Chem. C* **2014**, *118*, 2564–2569. [CrossRef]

24. Cheng, Y.; Wei, H.; Sun, R.; Tian, Z.; Zheng, X. Rapid Method for Protein Quantitation by Bradford Assay after Elimination of the Interference of Polysorbate 80. *Anal. Biochem.* **2016**, *494*, 37–39. [CrossRef] [PubMed]

25. Pecorelli, T.A.; Dibrell, M.M.; Li, Z.; Thomas, C.R.; Zink, J.I. Multifunctional inorganic nanoparticles for imaging, targeting, and drug delivery. *ACS Nano* **2008**, *2*, 889–896.

26. Angelos, S.; Liong, M.; Choi, E.; Zink, J.I. Mesoporous silicate materials as substrates for molecular machines and drug delivery. *Chem. Eng. J.* **2008**, *137*, 4–13. [CrossRef]

27. Michailidis, M.; Sorzabal-Bellido, I.; Adamidou, E.A.; Diaz-Fernandez, Y.A.; Aveyard, J.; Wengier, R.; Grigoriev, D.; Raval, R.; Benayahu, Y.; Shchukin, D.; et al. Modified Mesoporous Silica Nanoparticles with a Dual Synergetic Antibacterial Effect. *ACS Appl. Mater. Interfaces* **2017**, *9*, 38364–38372. [CrossRef]

28. Qin, L.; Feng, X.; Hafezi, M.; Zhang, Y.; Guo, J.; Dong, G.; Qin, Y. Investigating the Tribological and Biological Performance of Covalently Grafted Chitosan Coatings on Co-Cr-Mo Alloy. *Tribol. Int.* **2018**, *127*, 302–312. [CrossRef]

29. Cassie, A.B.D.; Baxter, S. Wettability of Porous Surfaces. *Trans. Faraday Soc.* **1944**, *40*, 546–550. [CrossRef]

30. Qin, L.; Lin, P.; Zhang, Y.; Dong, G.; Zeng, Q. Influence of Surface Wettability on the Tribological Properties of Laser Textured Co-Cr-Mo Alloy in Aqueous Bovine Serum Albumin Solution. *Appl. Surf. Sci.* **2013**, *268*, 79–86. [CrossRef]

31. Baron, A.; Quadrio, M.; Vigevano, L. On the boundary layer/riblets interaction mechanisms and the prediction of turbulent drag reduction. *Int. J. Heat Fluid Flow* **1993**, *14*, 324–332. [CrossRef]

32. Dean, B.; Bhushan, B. Shark-Skin Surfaces for Fluid-Drag Reduction in Turbulent Flow: A Review. *Philos. Trans. R. Soc. A Math. Phys. Eng. Sci.* **2010**, *368*, 5737. [CrossRef]

33. Jafargholinejad, S.; Pishevar, A.; Sadeghy, K. On the Use of Rotating-Disk Geometry for Evaluating the Drag-Reducing Efficiency of Polymeric and Surfactant Additives. *J. Appl. Fluid Mech.* **2011**, *4*, 1–5.

34. Pertsin, A.J.; Grunze, M. Computer Simulation of Water near the Surface of Oligo(ethylene glycol)-Terminated Alkanethiol Self-Assembled Monolayers. *Langmuir* **2000**, *16*, 8829–8841. [CrossRef]

35. Hou, J.; Dong, G.; Ye, Y.; Chen, V. Enzymatic degradation of bisphenol-A with immobilized laccase on TiO_2 solgel coated PVDF membrane. *J. Membr. Sci.* **2014**, *469*, 19–30. [CrossRef]

36. Zuo, G.; Wang, R. Novel membrane surface modification to enhance anti-oil fouling property for membrane distillation application. *J. Membr. Sci.* **2013**, *447*, 26–35. [CrossRef]

37. Hashino, M.; Hirami, K.; Ishigami, T.; Ohmukai, Y.; Maruyama, T.; Kubota, N.; Matsuyama, H. Effect of kinds of membrane materials on membrane fouling with BSA. *J. Membr. Sci.* **2011**, *384*, 157–165. [CrossRef]

38. Ponsonnet, L.; Reybier, K.; Jaffrezic, N.; Comte, V.; Lagneau, C.; Lissac, M.; Martelet, C. Relationship between surface properties (roughness, wettability) of titanium and titanium alloys and cell behaviour. *Mater. Sci. Eng. C* **2003**, *23*, 551–560. [CrossRef]

39. Cottin-Bizonne, C.; Barentin, C.; Bocquet, L. Scaling laws for slippage on superhydrophobic fractal surfaces. *Phys. Fluids* **2012**, *24*, 12001. [CrossRef]

40. Cao, M.; Li, Z.; Ma, H.; Geng, H.; Yu, C.; Jiang, L. Is Superhydrophobicity Equal to Underwater Superaerophilicity: Regulating the Gas Behavior on Superaerophilic Surface via Hydrophilic Defects. *ACS Appl. Mater. Interfaces* **2018**, *10*, 20995–21000. [CrossRef]

41. Fan, L.; Li, B.; Zhang, J. Antibioadhesive Superhydrophobic Syringe Needles Inspired by Mussels and Lotus Leafs. *Adv. Mater. Interfaces* **2015**, *2*, 1500019. [CrossRef]

Article

Fabrication and Characterization of Micro-Nano Electrodes for Implantable BCI

Ye Xi [1], Bowen Ji [1], Zhejun Guo [1], Wen Li [2,*] and Jingquan Liu [1,*]

[1] National Key Laboratory of Science and Technology on Micro/Nano Fabrication Laboratory, Department of Micro/Nano-electronics, Shanghai Jiao Tong University, Shanghai 200240, China; yesiki@sjtu.edu.cn (Y.X.); jibowen2015@sjtu.edu.cn (B.J.); gzj98762@sjtu.edu.cn (Z.G.)

[2] Department of Electrical and Computer Engineering, College of Engineering, Michigan State University, East Lansing, MI 48823, USA

* Correspondence: wenli@egr.msu.edu (W.L.); jqliu@sjtu.edu.cn (J.L.)

Received: 13 March 2019; Accepted: 8 April 2019; Published: 11 April 2019

Abstract: Signal recording and stimulation with high spatial and temporal resolution are of increasing interest with the development of implantable brain-computer interfaces (BCIs). However, implantable BCI technology still faces challenges in the biocompatibility and long-term stability of devices after implantation. Due to the cone structure, needle electrodes have advantages in the biocompatibility and stability as nerve recording electrodes. This paper develops the fabrication of Ag needle micro/nano electrodes with a laser-assisted pulling method and modifies the electrode surface by electrochemical oxidation. A significant impedance reduction of the modified Ag/AgCl electrodes compared to the Ag electrodes is demonstrated by the electrochemical impedance spectrum (EIS). Furthermore, the stability of modified Ag/AgCl electrodes is confirmed by cyclic voltammogram (CV) scanning. These findings suggest that these micro/nano electrodes have a great application prospect in neural interfaces.

Keywords: micro/nano electrode; electrochemistry; modified electrode; implantable BCI

---

1. Introduction

The multidisciplinary and multilevel comprehensive research on the higher cognitive function of the human brain has become one of the mainstream directions in contemporary scientific development [1–9]. Brain-computer interface (BCI) technology emerges in this condition which establishes a direct communication and control channel between the human brain and computers, or other electronic devices. Neuroscience studies show that electrical activity of the nervous system will have a corresponding change after the brain produces a motion of consciousness or receives certain outside stimulation. Through BCI technology, these neural signals can be recorded and converted into relevant command signals to drive external devices. One of the key aspects of BCI technology is the effective acquisition of physiological signals in neural activity through implantable electrodes [4]. Analyzing the bioelectrical signals collected by the electrodes will help us to comprehend brain mechanisms more clearly. Therefore, it is of great significance to develop implantable electrodes with novel structure and materials for BCI technology. For instance, Wang M. et al. successfully fabricated the microelectrodes modified with direct electrodeposition of composites for application in neural interfaces [8]. Wu, F. et al. prepared integrated μLEDs on silicon neural probes for a high-resolution optogenetic study [9]. However, the BCI technology still faces challenges in the biocompatibility and long-term implantation of devices after implantation [10–14]. Because of the mechanical mismatch between the material and the nerve tissue, the electrodes will be encapsulated by glial cells due to the chronic foreign body response after long-term implantation. Therefore, the electrode recording points will lose the function of accurately recording electrical signals. In addition, the device implantation can cause damage to the neurons which are in the same plane with, or near the recording points of the electrode, and the reliability of the measured bioelectrical signals will be also affected. Compared to electrodes with a structure similar to a flat plate, the damage to peripheral neurons caused by needle electrode implantation mainly occurs on the cone surface, which is not in the same plane with the electrode recording points. At the same time, the conical structure of needle electrodes can reduce damage to human tissue during the implantation [15,16].

A widespread method to fabricate needle micro/nano electrodes is the laser-assisted pulling method because of the excellent control of the electrode geometry and its high repeatability [17–24]. In this method, a tension is applied to both ends of the sample which consists of a glass tube and a metal wire inside it during the laser heating. During the process of laser heating and pulling, the plastic sample can be pulled down to a pair of independent metal wire tips trapped in glass with a smaller diameter. Then the tip is polished to expose the metal core. Except for the area of the electrode at the tip, the entire electrode is encased in a thin layer of glass. Multiple electrode recording points can be realized by pulling multi-barrel glass tubes, and the electrode array can also be realized by pulling multiple glass tubes together [23,24]. It can realize a multifunctional integration of electrodes through carbon pyrolysis, electroplating and modification of functional groups.

At present, common metal materials used in the laser-assisted pulling method mainly include Au and Pt. Due to the mismatch in physical properties, it is still difficult to prepare needle electrodes with metals other than Au and Pt [20]. Therefore, it is of great significance to develop a laser-assisted pulling method with other materials. In addition to excellent ductility and the highest electrical and thermal conductivity of all metals, silver also has good prospects in the medical field. Silver is widely used in biomedical hydrogels, bandages and long-term implanted devices because of its antibacterial properties and biocompatibility [25,26]. In this paper, the Ag needle micro/nano electrodes were fabricated. By optimizing preparation process and parameters, the mismatch of melting point and other physical properties between quartz and silver were successfully resolved, and the Ag/AgCl needle microelectrodes were fabricated by electrolytic chlorination. The Ag/AgCl electrode is close to the non-polarized electrode at low current and the current flowing, through the electrode during bioelectrical signal measurement is weak. Therefore, Ag/AgCl electrodes are suitable for the determination of bioelectrical signals as a detection electrode [27,28]. The electrochemical impedance spectrum (EIS) method and stability test were carried out to characterize the Ag/AgCl electrodes after modification. Finally, the preparation of a needle nano electrode is discussed, on the basis of the preparation of needle micro electrodes. The SEM images confirmed the successful preparation of the Ag needle nanoelectrode.

2. Materials and Methods

2.1. Materials

Polyethylene glycol (PEG, 2000), potassium chloride (KCl), hydrochloric acid (HCl), acetone (CH_3COCH_3) and phosphate buffered saline (PBS, pH 7.4) were purchased from Sinopharm Chemical Reagent Company, Ltd. (Shanghai, China). All chemicals used in the experiments were of reagent grade quality or better and were used without further purification. Aqueous solutions were prepared from de-ionized water. Quartz capillaries (o.d. = 1.0 mm, i.d. = 0.3mm) were purchased from Sutter Instrument Company (Novato, CA, USA).

2.2. Instruments

The laser-based micropipette puller system (P-2000, Sutter Instrument Company) was used in the fabrication of the Ag electrodes. High vacuum scanning electron microscopy (ULTRA55, Zeiss, Jena, Gemany) was used for SEM observation. The anodic oxidation process of Ag electrodes was carried out on a CHI 660c electrochemical workstation (CH Instruments, Austin, TX, USA). Other electrochemical measurements such as cyclic voltammogram (CV) and electrochemical impedance spectrum (EIS) were conducted by Autolab (PGSTAT204, Metrohm, Herisau, Switzerland). An electrode beveler (BV-10, Sutter Instrument Company) was used in the polishing process.

2.3. Preparation of the Ag Needle Electrode

Figure 1 shows a five-step fabrication process of Ag needle electrodes. At first, silver wires were pretreated with fine grit sandpapers with different grits to remove oxides from the surface. Afterward they were treated with ultrasonic cleaning in acetone, ethanol and de-ionized water and dried in a drying oven at 200° Celsius, a ~2 mm long piece of silver wire was inserted into a 75 mm long piece of quartz capillary (o.d. = 1.0 mm, i.d. = 0.3mm). It is worth noting that the length of silver wire will directly influence the morphology of the needle electrode obtained by the laser-assisted pulling method under certain optimization parameters.

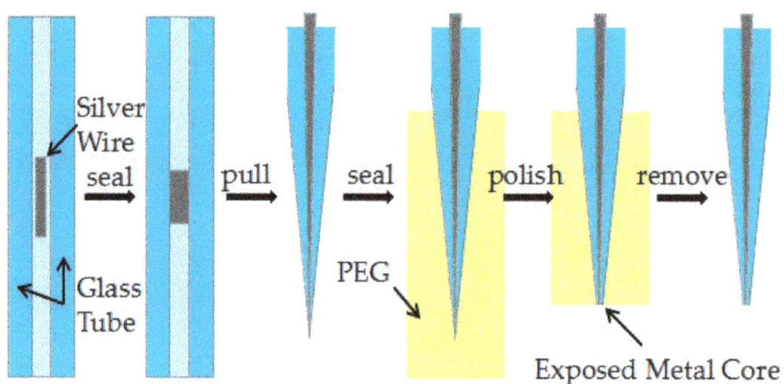

Figure 1. Process diagram showing the preparation of needle electrodes with the laser-assisted pulling method.

In the second step, the silver wire was sealed in quartz glass by a butane torch. The temperature of the butane torch flame was about 1300° Celsius, which is between the melting temperature of the silver wire and the softening temperature of quartz glass. Heated by the butane torch, the silver melted and then adhered to the inner wall of the glass tube. As shown in Figure 2a, a rotating fixture platform was adopted in the melting process in order to obtain a better adhesion between the silver wire and quartz glass inner wall. After the melting process, the sealing was checked under an optical microscope to ensure that silver wire adhered to the inner wall well.

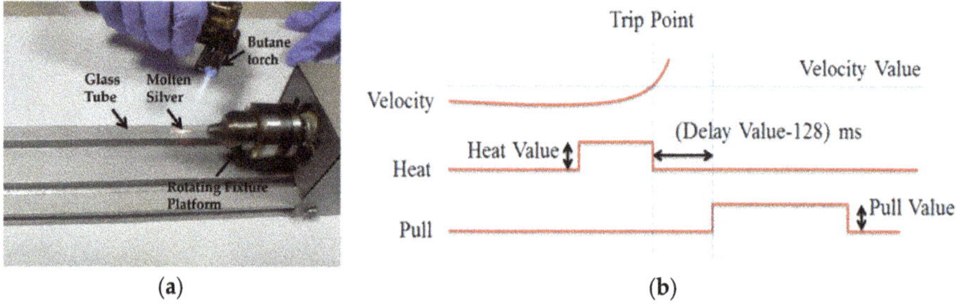

(a)   (b)

Figure 2. (**a**) Picture of the rotating fixture platform adopted in the melting process. (**b**) A typical pull cycle based on the P-2000 laser-based micropipette puller system.

In the third step, a laser-assisted pulling method was adopted to prepare the Ag needle electrodes. The P-2000 laser-based micropipette puller system was used in the pulling process. Figure 2b shows a typical pull cycle based on the P-2000 laser-based puller system. The adjustment parameters of the laser needle puller system were as follows: Heat, Velocity, Delay, Pull and Filament. The parameter Heat (range from 0 to 999) controls the power of the laser applied on the capillary glass tube. The typical range of Heat setting is around 700 to 999 for quartz. The parameter Velocity (range from 0 to 255) is a temperature-dependent parameter that determines the trip point at which the equipment stops laser heating. The sample will have a higher softening degree before hard pulling with an increase of the parameter Velocity. The parameter Delay (range from 0 to 255) determines the time interval between stopping laser heating and applying a pull. A change of one unit represents a change of 1 ms to the delay time. When the value of parameter Delay is 128, the equipment will stop heating and apply tension at the same time. If the parameter Delay is larger than 128, the hard pull will begin after the end of a laser heating. If the parameter Delay is smaller than 128, the hard pull will begin before the end of a laser heating. The parameter Pull (range from 0 to 255) determines the magnitude of tension applied at both ends of the glass tube. In general, the higher the parameter Pull, the smaller the tip of electrode will be prepared. The parameter Filament (range from 0 to 5) specifies the scan length of the laser beam which is used to heat the designated area of the glass tube. When the parameter Filament is set from 0 to 5, the corresponding scanning lengths are 1 mm, 1.5 mm, 1.9 mm, 4.5 mm, 6.5 mm and 8 mm, respectively. These five parameters influence and restrict each other. The sealing, containing quartz glass and silver wire, was pulled into two independent silver tips trapped in quartz glass under certain parameters. In this step, the morphology of the prepared needle electrode is significantly affected by different parameters.

In the fourth step, the tip portion was covered in PEG which is water-soluble. This operation was used to increase the rigidity of the ultrafine tip for subsequent polishing. The sample was roughly polished with sandpapers of different grits. By using a KCl solution containing alumina suspension on a polishing cloth, the silver tip was further exposed and sanded with the aid of a Sutter BV-10 electrode beveler which was situated just below a vertical micro-mobile platform. During the further polishing process, the sample was fixed on a micro-mobile platform with a displacement accuracy of 1 μm and the sample was polished with a decrease in the vertical direction. To control the exposure of the tip, the electrical resistance between the silver wire embedded in the composite material and the polishing cloth was tested after every certain depth of polishing through a multimeter [22]. Finally, the PEG wrapped around the outside of the entire tip was removed in hot liquids and a Ag electrode with a smooth surface was obtained.

2.4. Modification of the Ag Needle Electrode

The Ag electrode was cleaned with acetone and de-ionized water to remove surface contaminants. Then the electrode was anodized in 0.1 M HCl aqueous solution on a CHI 660c electrochemical workstation. The platinum electrode was used as a cathode and the Ag electrode was connected to the anode. The whole anodic oxidation process was under a constant voltage of 1 V.

3. Results and Discussion

3.1. Fabrication of the Ag Needle Microelectrode

Silver has a melting point of 961.78° Celsius which is far below the softening temperature of quartz glass. In the process of sealing the silver into the quartz glass tube with a butane torch, the silver melted first, while the quartz glass tube remained solid. The melted silver solidified and adhered to the inner wall of the quartz glass tube as the temperature decreased. The attached quality between silver and the inner wall of the glass tube would affect the morphology of prepared electrode. Figure 3a shows the adhesion surface between the silver and the inner wall of the glass tube. The adhesion surface was dotted with some defects, which may lead to the discontinuity of the internal silver wire as shown in Figure 3b. In order to improve the adhesion quality, a rotating fixture platform was adopted in the heating melting process. In the high-speed rotation, the silver was tightly attached to the inner wall of the glass tube. Through analysis and optimization of various parameters in laser-based puller system, the Ag needle electrode was successfully fabricated under the following parameters:

Heat = 850, Filament = 4, Velocity = 30, Delay = 150 and Pull = 170. As shown in Figure 4, the results confirmed that the mismatch between quartz and silver was successfully resolved by optimizing preparation process and parameters. The effective recording point of the electrode tip was ~7.9 μm in diameter and the variable-diameter part of the needle electrode was ~9.2 mm in length. Micrographs showed that the needle electrode presented a good internal continuity and there was no cracks in the surrounding glass.

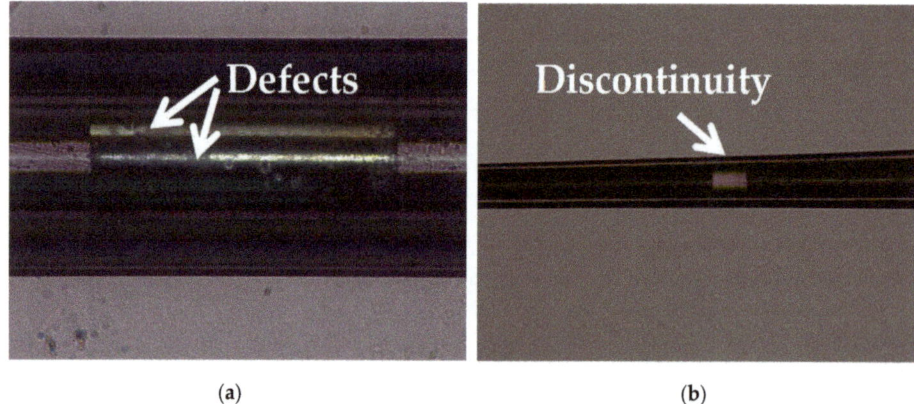

(a)   (b)

Figure 3. (a) Micrograph of the adhesion surface between the silver and the inner wall of the glass tube. (b) Micrograph of the discontinuity occurred in the internal silver wire.

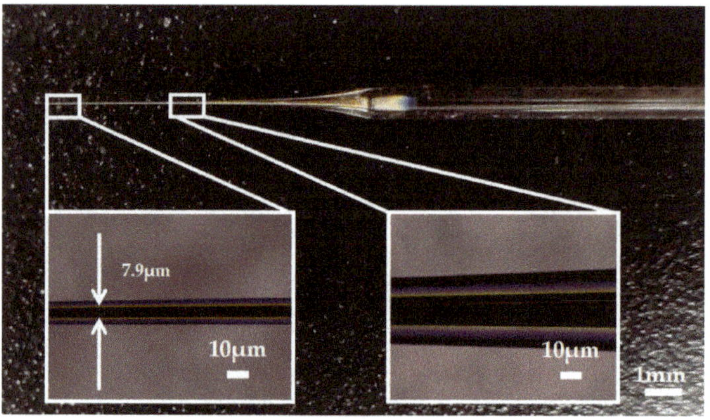

Figure 4. The picture of the Ag needle electrode.

3.2. Electrochemical Characterization of the Needle Microelectrodes

By an anodic oxidation process, silver atoms were oxidized to silver ions at the electrode-electrolyte interface and then deposited on the electrode surface after combining with chloride ions. Equation (1) and (2) express the specific reaction process occurring in the anodic oxidation process [28]:

$$Ag \leftrightarrow Ag^+ + e \tag{1}$$

$$Ag^+ + Cl^- \leftrightarrow AgCl \downarrow \tag{2}$$

As seen in Figure 5, the impedance of the modified Ag/AgCl electrode decreased compared to the Ag electrode at a low frequency from 0.1 Hz to 1000 Hz. The impedances of electrodes at low frequency are compared in Table 1. This experiments result was in agreement with the theory that silver chloride is considered as a type of non-polarizable material while silver is a polarizable material. Based on the excellent electrical conductivity of silver, the impedance of the modified Ag/AgCl electrode further reduced at low frequency and it was advantageous for measuring and recording bio-electricity signals.

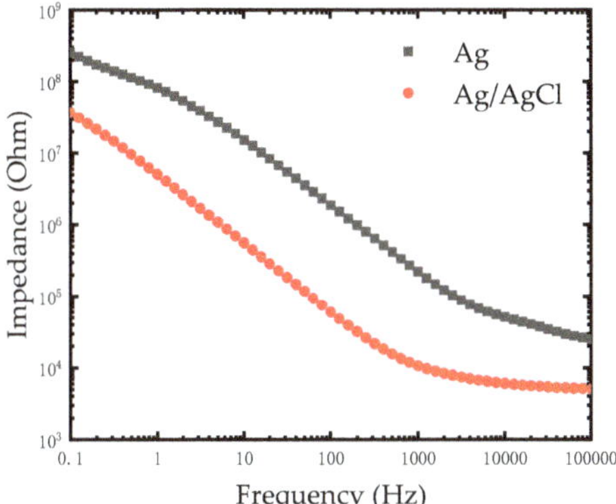

Figure 5. Impedance comparison diagram of the Ag electrode and modified Ag/AgCl electrode in phosphate buffered saline (PBS) solution.

Table 1. The impedances of the Ag electrode and modified Ag/AgCl electrode at different frequencies.

Frequency (Hz)	The Ag Electrode (Ω)	The Ag/AgCl Electrode (Ω)
1000	2.2×10^5	1.05×10^4
100	1.87×10^6	6.01×10^4
10	1.52×10^7	5.56×10^5
1	8.19×10^7	5.01×10^6
0.1	2.56×10^8	3.69×10^7

3.3. Stability Tests

After chlorination of Ag electrodes, a thin film of silver chloride was formed on the surface of the electrodes. The quality of the film would directly affect the stability of electrodes. The stability of the modified electrodes was evaluated by repetitive CV scanning. The CV scanning process was operated in PBS solution at room temperature. The Ag/AgCl electrode and the Pt electrode were used as anodes and cathodes respectively. We carried out 200 cycles of CV scanning with the scan rate of 1 V/s between -0.6 V and 0.8 V and the EIS of electrodes was measured. As shown in Figure 6a, the impedance of electrodes remained relatively stable after 200 cycles. The impedance of the electrode at 1 KHz was reduced by one fifth (from 9140.11 Ω to 7312.46 Ω). The reduction of impedance may be caused by further chlorination of silver on the electrode surface after rapid CV scanning. This is because the reaction of the Ag/AgCl electrode is an electrolytic reaction rather than a polarization of the metal. The following equilibrium reaction occurs on the electrode surface:

$$AgCl + e \leftrightarrow Ag + Cl^- \tag{3}$$

Therefore, the thickness of the AgCl layer changes periodically during the rapid CV scanning. The results in Figure 6b shows the phase of electrodes after 200 cycles. Considering that the typical electroencephalography (EEG) signal amplitude was less than 1 mV, the rapid CV scanning test equaled a destructive test to detect the stability of the Ag/AgCl electrode. The experimental results demonstrated that the modified Ag/AgCl electrodes performed with good stability, which could meet the neural signal measurement in a complex environment in vivo.

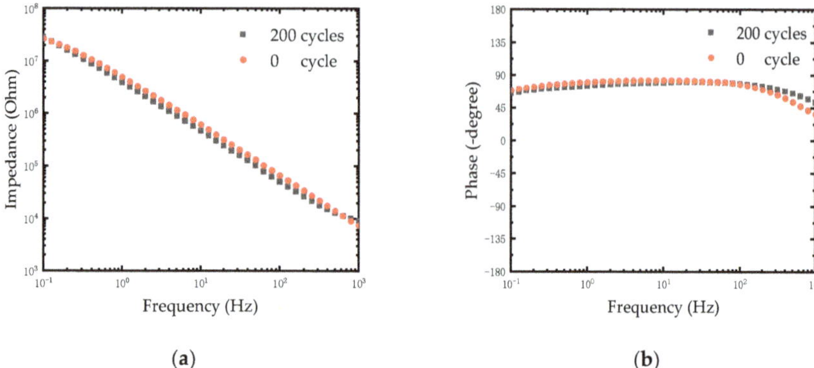

(a)   (b)

Figure 6. (**a**) Impedance plot of modified Ag/AgCl electrode treated with different cycles of CV scanning. (**b**) Phase spectra of modified Ag/AgCl electrode treated with different cycles of CV scanning.

3.4. SEM Images Characterization of the Ag Needle Nanoelectrodes

For microelectrode preparation, we fabricated the needle electrodes at nano-scale which had the ability to record signals in a single neuron cell due to its great spatial resolution. SEM characterization was carried out to observe the morphology of the Ag needle submicron electrodes. Figure 7a shows a SEM image of an entire silver/glass tip after the pulling process under the following parameters: Heat = 910, Filament = 2, Velocity = 30, Delay = 128 and Pull = 250. Compared to the parameters in the preparation of the microelectrodes, the increase of laser power and the decrease of scan length play significant roles in the reduction of electrode size. A SEM image with an enlarged scale of a silver/glass tip is shown in Figure 7b. Because the diameter ratio between the quartz and the silver wire was consistent with the initial ratio between the quartz tube and the silver wire [21], the diameter of the Ag electrode was less than 100 nm. This result confirmed a successful preparation of Ag needle electrodes at the nano-scale.

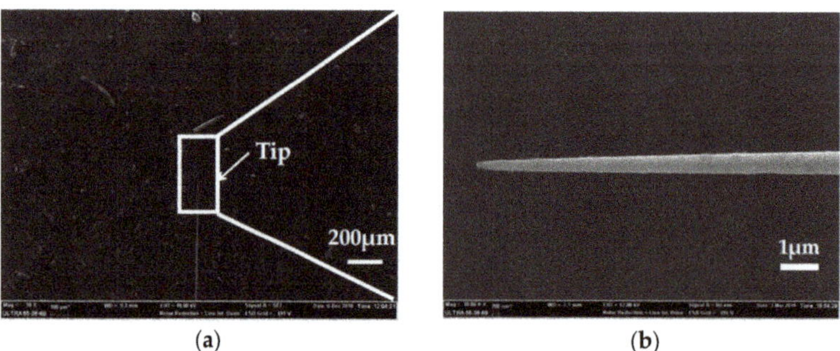

(a)   (b)

Figure 7. (**a**) SEM image of the entire silver/glass tip of the Ag nanoelectrode. (**b**) SEM image with an enlarged scale of the silver/glass tip of the Ag nanoelectrode.

4. Conclusions

In conclusion, we developed the fabrication of Ag needle micro/nano electrodes with a laser-assisted pulling method. Through analyzing and optimizing the parameters in laser-based puller system, the Ag needle microelectrodes with a diameter of 7.9 μm were successfully fabricated. An anodic oxidation process which converted silver to silver chloride was operated to modify the electrode surface. The electrochemistry performance of the modified Ag/AgCl electrode was researched by studying the EIS. After modification, the impedance of electrode at 1 KHz changed from 2.2×10^5 Ω to 1.05×10^4 Ω.

The results showed that the impedance of the modified Ag/AgCl electrodes presented a significant reduction after surface modification. The stability of modified Ag/AgCl electrodes was also verified by CV scanning. On the basis of the microelectrode preparation, we prepared the needle electrodes at the nano-scale which can further improve the spatial resolution of the device. These findings suggest that these micro/nano electrodes have great application prospects in neural interfaces. In future work, we will continue to focus on electrode scaling reduction and single cell signal acquisition.

Author Contributions: Data curation, Y.X.; Investigation, Y.X., B.J. and Z.G.; Supervision, W.L. and J.L.; Writing—original draft, Y.X., B.J. and Z.G.; Writing—review & editing, W.L. and J.L.

Funding: This work was partially funded by the National Natural Science Foundation of China (grant number 61728402), the Research Program of Shanghai Science and Technology Committee (grant number 17JC1402800), the Program of Shanghai Academic/Technology Research Leader (grant number 18XD1401900). The authors are also grateful to the Center for Advanced Electronic Materials and Devices (AEMD) of Shanghai Jiao Tong University.

Conflicts of Interest: The authors declare no conflict of interest.

References

1. Jun, J.J.; Steinmetz, N.A.; Siegle, J.H.; Denman, D.J.; Bauza, M.; Barbarits, B.; Lee, A.K.; Anastassiou, C.A.; Andrei, A.; Aydın, Ç.; et al. Fully integrated silicon probes for high-density recording of neural activity. *Nature* **2017**, *551*, 232–236. [CrossRef]

2. Canales, A.; Jia, X.; Froriep, U.P.; Koppes, R.A.; Tringides, C.M.; Selvidge, J.; Lu, C.; Hou, C.; Wei, L.; Fink, Y.; et al. Multifunctional fibers for simultaneous optical, electrical and chemical interrogation of neural circuits in vivo. *Nat. Biotechnol.* **2015**, *33*, 277–284. [CrossRef]

3. Rivnay, J.; Wang, H.; Fenno, L.; Deisseroth, K.; Malliaras, G.G. Next-generation probes, particles, and proteins for neural interfacing. *Sci. Adv.* **2017**, *3*, e1601649. [CrossRef] [PubMed]

4. Alivisatos, A.P.; Chun, M.; Church, G.M.; Greenspan, R.J.; Roukes, M.L.; Yuste, R. The brain activity map project and the challenge of functional connectomics. *Neuron* **2012**, *74*, 970–974. [CrossRef] [PubMed]

5. Green, R.; Abidian, M.R. Conducting Polymers for Neural Prosthetic and Neural Interface Applications. *Adv. Mater.* **2015**, *27*, 7620–7637. [CrossRef] [PubMed]

6. Lee, H.J.; Son, Y.; Kim, J.; Lee, C.J.; Yoon, E.-S.; Cho, I.-J. A multichannel neural probe with embedded microfluidic channels for simultaneous in vivo neural recording and drug delivery. *Lab Chip* **2015**, *15*, 1590–1597. [CrossRef]

7. Normann, R.A.; Fernandez, E. Clinical applications of penetrating neural interfaces and Utah Electrode Array technologies. *J. Neural Eng.* **2016**, *13*, 61003. [CrossRef]

8. Wang, M.; Ji, B.; Gu, X.; Tian, H.; Kang, X.; Yang, B.; Wang, X.; Chen, X.; Li, C.; Liu, J. Direct electrodeposition of Graphene enhanced conductive polymer on microelectrode for biosensing application. *Biosens. Bioelectron.* **2018**, *99*, 99–107. [CrossRef]

9. Wu, F.; Stark, E.; Ku, P.-C.; Wise, K.D.; Buzsáki, G.; Yoon, E. Monolithically integrated μLEDs on silicon neural probes for high-resolution optogenetic studies in behaving animals. *Neuron* **2015**, *88*, 1136–1148. [CrossRef] [PubMed]

10. Geon, K.; Sung, W.L.; Hee, C.L.; Il-Joo, C.; Hyunjoo, J.L. Neural Probes for Chronic Applications. *Micromachines* **2016**, *7*, 179. [CrossRef]

11. Pei, W. Development of implantable silicon neural microelectrodes. *Sci. Technol. Rev.* **2018**, *6*, 77–82.

12. Seymour, J.P.; Kipke, D.R. Neural probe design for reduced tissue encapsulation in CNS. *Biomaterials* **2007**, *28*, 3594–3607. [CrossRef] [PubMed]

13. Grand, L.; Wittner, L.; Herwik, S.; Göthelid, E.; Ruther, P.; Oscarsson, S.; Neves, H.; Dombovári, B.; Csercsa, R.; Karmos, G. Short and long term biocompatibility of neuroprobes silicon probes. *J. Neurosci. Methods* **2010**, *189*, 216–229. [CrossRef]

14. Karumbaiah, L.; Saxena, T.; Carlson, D.; Patil, K.; Patkar, R.; Gaupp, E.A.; Betancur, M.; Stanley, G.B.; Carin, L.; Bellamkonda, R.V. Relationship between intracortical electrode design and chronic recording function. *Biomaterials* **2013**, *34*, 8061–8074. [CrossRef] [PubMed]

15. Boutte, R.W.; Merlin, S.; Yona, G.; Griffiths, B.; Angelucci, A.; Kahn, I.; Shoham, S.; Blair, S. Utah optrode array customization using stereotactic brain atlases and 3-D CAD modeling for optogenetic neocortical interrogation in small rodents and nonhuman primates. *Neurophotonics* **2017**, *4*, 41502. [CrossRef] [PubMed]

16. Richardson, A.G.; Weigand, P.K.; Sritharan, S.Y.; Lucas, T.H. A chronic neural interface to the macaque dorsal column nuclei. *J. Neurophysiol.* **2016**, *115*, 2255–2264. [CrossRef] [PubMed]

17. Clausmeyer, J.; Schuhmann, W. Nanoelectrodes: Applications in electrocatalysis, single-cell analysis and high-resolution electrochemical imaging. *TrAC Trends Anal. Chem.* **2016**, *79*, 46–59. [CrossRef]

18. Pendley, B.D.; Abruna, H.D. Construction of submicrometer voltammetric electrodes. *Anal. Chem.* **1990**, *62*, 782–784. [CrossRef]

19. Fish, G.; Bouevitch, O.; Kokotov, S.; Lieberman, K.; Palanker, D.; Turovets, I.; Lewis, A. Ultrafast response micropipette-based submicrometer thermocouple. *Rev. Sci. Instrum.* **1995**, *66*, 3300–3306. [CrossRef]

20. Hua, H.; Liu, Y.; Wang, D.; Li, Y. Size-Dependent Voltammetry at Single Silver Nanoelectrodes. *Anal. Chem.* **2018**, *90*, 9677–9681. [CrossRef]

21. Li, Y.; Bergman, D.; Zhang, B. Preparation and Electrochemical Response of 1–3 nm Pt Disk Electrodes. *Anal. Chem.* **2009**, *81*, 5496–5502. [CrossRef]

22. Zhang, B.; Galusha, J.; Shiozawa, P.G.; Wang, G.; Bergren, A.J.; Jones, R.M.; White, R.J.; Ervin, E.N.; Cauley, C.C.; White, H.S. Bench-Top Method for Fabricating Glass-Sealed Nanodisk Electrodes, Glass Nanopore Electrodes, and Glass Nanopore Membranes of Controlled Size. *Anal. Chem.* **2007**, *79*, 4778–4787. [CrossRef] [PubMed]

23. Lin, Y.; Trouillon, R.; Svensson, M.I.; Keighron, J.D.; Cans, A.; Ewing, A.G. Carbon-Ring Microelectrode Arrays for Electrochemical Imaging of Single Cell Exocytosis: Fabrication and Characterization. *Anal. Chem.* **2012**, *84*, 2949–2954. [CrossRef]

24. Xue, L.; Cadinu, P.; Paulose Nadappuram, B.; Kang, M.; Ma, Y.; Korchev, Y.; Ivanov, A.P.; Edel, J.B. Gated Single-Molecule Transport in Double-Barreled Nanopores. *ACS Appl. Mater. Interfaces* **2018**, *10*, 38621–38629. [CrossRef]

25. Slenters, T.V.; Hauser-Gerspach, I.; Daniels, A.U.; Fromm, K.M. Silver coordination compounds as light-stable, nano-structured and anti-bacterial coatings for dental implant and restorative materials. *J. Mater. Chem.* **2008**, *18*, 5359–5362. [CrossRef]

26. Fordham, W.R.; Redmond, S.; Westerland, A.; Cortes, E.G.; Walker, C.; Gallagher, C.; Medina, C.J.; Waecther, F.; Lunk, C.; Ostrum, R.F.; et al. Silver as a Bactericidal Coating for Biomedical Implants. *Surf. Coat. Technol.* **2014**, *253*, 52–57. [CrossRef]

27. Peng, H.L.; Liu, J.Q.; Dong, Y.; Yang, B.; Chen, X.; Yang, C. Parylene-based flexible dry electrode for bioptential recording. *Sens. Actuators B Chem.* **2016**, *231*, 1–11. [CrossRef]

28. Xu, P.J.; Liu, H.; Zhang, H.; Tao, X.M.; Wang, S.Y. Electrochemical Modification of Silver Coated Multifilament for Wearable ECG Monitoring Electrodes. *Adv. Mater. Res.* **2011**, *332*, 1019–1023. [CrossRef]

Article

Single-Sided Near-Field Wireless Power Transfer by A Three-Dimensional Coil Array

Amirhossein Hajiaghajani [1] and Seungyoung Ahn [2,*]

[1] Department of Electrical Engineering and Computer Science, University of California, Irvine, CA 92697, USA; ahajiagh@uci.edu
[2] CCS Graduate School for Green Transportation, Korea Advanced Institute of Science and Technology (KAIST), Daejeon 305-701, Korea
* Correspondence: sahn@kaist.ac.kr

Received: 13 February 2019; Accepted: 19 March 2019; Published: 21 March 2019

Abstract: Wirelessly powered medical microrobots are often driven or localized by magnetic resonance imaging coils, whose signal-to-noise ratio is easily affected by the power transmitter coils that supply the microrobot. A controlled single-sided wireless power transmitter can enhance the imaging quality and suppress the radiation leakage. This paper presents a new form of electromagnet which automatically cancels the magnetic field to the back lobes by replacing the traditional circular coils with a three-dimensional (3D) coil scheme inspired by a generalized form of Halbach arrays. It is shown that, along with the miniaturization of the transmitter system, it allows for improved magnetic field intensity in the target side. Measurement of the produced magnetic patterns verifies that the power transfer to the back lobe is 15-fold smaller compared to the corresponding distance on the main lobe side, whilst maintaining a powering efficiency similar to that of conventional planar coils. To show the application of the proposed array, a wireless charging pad with an effective powering area of 144 cm^2 is fabricated on 3D-assembled printed circuit boards. This 3D structure obviates the need for traditional magnetic shield materials that place limitations on the working frequency and suffer from non-linearity and hysteresis effects.

Keywords: biomedical microrobot; magnetic devices; magnetic shielding; wireless power transfer

---

1. Introduction

Wireless power transfer (WPT) plays a key role in developing instruments that rely on external powering such as electric vehicles [1], micro-devices [2], and biomedical implants [3,4]. Over the past decade, new designs for coil structures have expanded the usage of WPT in various applications. Early wireless powering structures were based on the resonant coupling of aligned coils [5] and geometry optimization of the power transmitter (Tx) and receiver (Rx) coils [6]. Such structures generate intense magnetic fields over the surrounding space and are able to cause electromagnetic (EM) interference with nearby electric equipment [7,8]. In addition to non-thermal effects [9], an unintentional EM leakage can critically threaten several biomedical procedures such as electric pacemakers [10], magnetic drug targeting [11,12], and magnetic innervation [13]. In addition, several recent techniques for microrobot technology rely on WPT coils in combination with magnetic resonance imaging (MRI) coils for real-time feedback on the function and location of the microrobot [14–18]. However, the magnetic actuators responsible for wireless powering to microrobots often consist of pairs of Maxwell and Helmholtz coils that can easily affect the signal-to-noise ratio in the MRI's radio-frequency pickup coils [19–21]. A single-sided WPT actuator that only affects the microrobot's operation region and reduces the interference with the magnetic field of the MRI's pickup coils can be a solution to this challenge (see Figure 1).

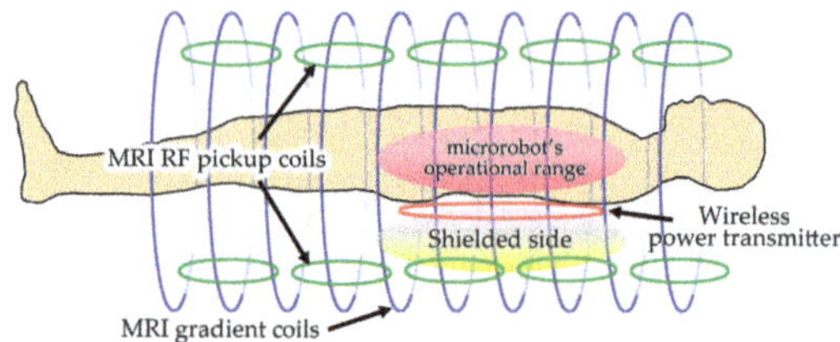

Figure 1. Schematic of the simultaneous usage of microrobot actuation and magnetic resonance imaging (MRI) coils to provide function and location feedback for implanted medical devices.

Moreover, in several applications such as wireless powering to biomedical devices [22,23], the Tx coils lie in the vicinity of lossy media such as living tissues. Conventional Tx coils create magnetic fields on either side of the transmitter, induce unintentional electric currents on the tissues, and have the potential to cause thermal damage [24]. Whilst magnetic sheets have been used to design shields to suppress the EM leakage in such conductive media [25], such techniques have limited the working frequency due to their non-linear behavior as well as complicated the cooling systems [26]. Generating a spatially controlled magnetic pattern for WPT that minimizes the risk of EM interference between multiple Rx coils has been a challenge for many biomedical systems.

To this aim, the spatial forming of subwavelength magnetic fields is a potential solution and may enable the control of the produced magnetic field and the formation of a single-sided radiation pattern aimed only at the Rx coils and away from the lossy media [12,27–29]. A traditional method to focus the magnetic field and suppress its leakage has been introduced by Halbach [30], which has since been the basis of a multitude of biomedical treatments such as magnetic drug delivery [31–33]. However, this scheme is based on arrays of permanent magnets and cannot be directly used for generating alternating magnetic fields. In a previous study, it was demonstrated that a combination of multi-turn coils in the form of linear Halbach arrays had the potential for lowering the EM leakage [34]. However, since bulky multi-turn coils generate enormous amounts of heat, such a structure cannot be used in compact designs such as charging pads and untethered powering to microrobots.

This paper is the first time a set of 3D coil arrays, based on the Halbach arrays and using vertical and horizontal circuit boards, has been demonstrated to improve the efficacy of low-leakage WPT using a single-sided radiation pattern. We aimed to minimize the magnetic field produced on the shielded side and reinforce it on the other side. The efficiency of the proposed design was validated through experimental tests at different distances from the Tx structure. The goal of this work was to investigate the practical potentiality of 3D coil arrays for low-leakage wireless powering. The total efficiency of the power transfer depends on the microrobot's geometry, Rx coil design, and distance from the Tx. Compared to conventional wireless charging pads, the magnetic fields were blocked on one side of the introduced array of coils to lower the risk of EM interference.

2. Evolution of Halbach Arrays

Halbach arrays allow magnetic flow to be formed only on one side and weakens the magnetic field's intensity on the opposing side. Generally, a coil-based scheme can substitute for a linear array of permanent magnets by considering the magnetization vector of each element and ensuring that the new coils generate a magnetic potential similar to the magnets. Based on the Halbach arrays, a one-dimensional (1D) coil scheme was introduced [34] (see Figure 2b), though it is ineligible for miniaturized wireless charging pads due to its bulky and difficult-to-fabricate structure. It also generates cross-polarized magnetic fields whose magnetic energy cannot be harvested even by a well-aligned Rx coil.

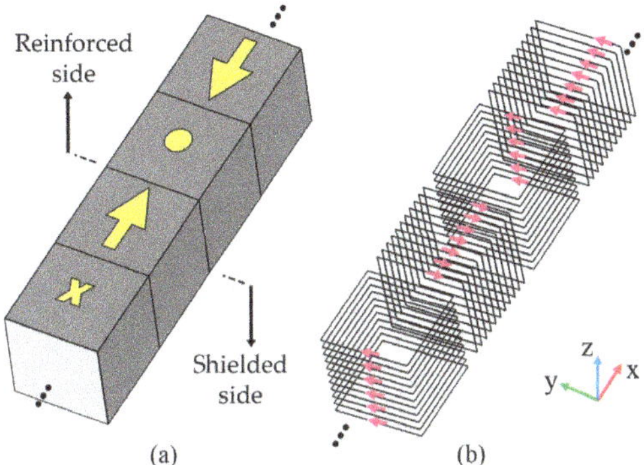

Figure 2. (a) A typical linear Halbach array in which the magnetization vectors are noted and (b) its coil-based equivalent.

To reduce the cross-polarized fields and lower the overall weight of the Tx, the conventional linear Halbach chain was generalized to a planar two-dimensional (2D) array by adding perpendicular current paths. The schematic of the current in the unit element is shown in Figure 3, where the adjacent elements run currents in opposite directions. The magnetic near field of the path element running a current of I (phasor notation) is derived by [35]:

$$\vec{H}_c(x,y,z) = \frac{I}{4\pi} \int_c \frac{\vec{dl'} \times \left(\vec{r} - \vec{r'} \right)}{\left| \vec{r} - \vec{r'} \right|^3} \tag{1}$$

where r and r' denote the vectors of the observation point and the point on the current path element c (shown in Figure 3).

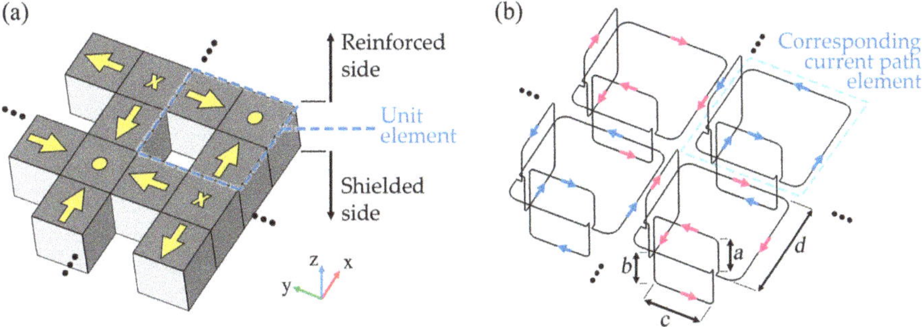

Figure 3. The geometry of the introduced 3D structure that represents an element of the planar array. The array may be extended in both x and y directions.

The total magnetic field produced by curves with an array size of $M \times N$ (in x and y directions, respectively) is obtained by the superposition principle:

$$\vec{H}_{total} = \sum_{m=0}^{M-1} \sum_{n=0}^{N-1} (-1)^{n+m} \vec{H}_c(x - nd,\ y - md, z). \tag{2}$$

The maximum side length of d depends on the working frequency such that the entire array's size becomes significantly smaller than the wavelength. The minimum d is approximately equal to the distance between the Rx and Tx coils based on the analytical solution of magnetic fields generated by subwavelength coils [36]. The lateral loops are attached symmetrically (i.e., $a = b$, see Figure 3b).

The entire array scheme was modeled using the software COMSOL and the resultant spatial pattern of magnetic fields on a typical *y*-plane is shown in Figure 4b. It was observed that although no shielding material was employed, the main lobes of the magnetic fields take place only above the structure and play the role for power transfer. Meanwhile, the minor back lobes made by the lateral coils are unable to form an intense magnetic field or transfer power to nearby electric facilities under the Tx coils. Figure 4c compares the proposed single-sided magnetic pattern with the traditional planar 2D coils (i.e., the same structure without the lateral coils).

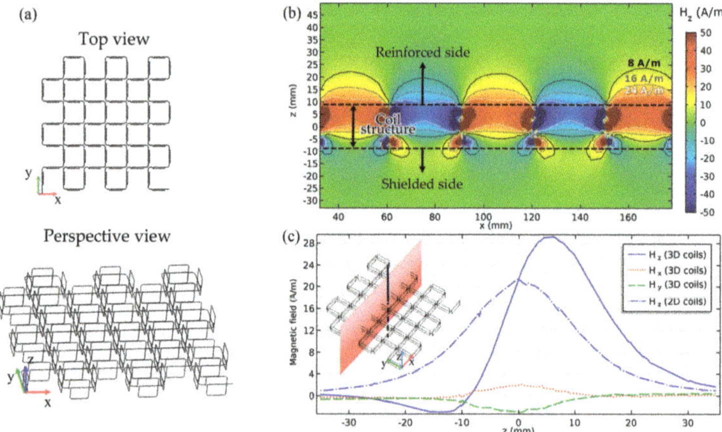

Figure 4. (**a**) Top and perspective views of the entire array scheme. (**b**) The intensity of the *z* component of magnetic fields in terms of phasors on the plane $y = 10.5$ cm for $a = b = 6$ mm, $d = 30$ mm, and $I = 1$ A. (**c**) A comparison between the proposed 3D coils and similar traditional 2D coils along the main lobe's axis at $x = y = 10.5$ cm and $I = 1$ A. $z < -7$ mm and $z > 7$ mm indicate the shielded and strengthened zones, respectively. -7 mm $< z < 7$ mm is an inaccessible region where the structure is located.

As shown in Figure 4b, the magnetic field's intensity peaks at $z = a$ while a near-zero field was observed at $z = -b$. Compared to traditional 2D actuators, the microrobot would receive a greater amplitude of magnetic fields above the structure (at $z > a$ and running the same actuating current). By placing the actuator between the body and the pickup coils of the MRI, it was observed that the magnetic field's intensity significantly decreases, which leads to minimizing the interference in the imaging coils. In addition, since the *x* and *y* components of the magnetic fields generated by the lateral coils cancel out each other in the main lobes, the cross polarization is insignificant.

3. Experimental Validation

In practice, the horizontal 2D structure was fabricated on a printed circuit board (PCB). Long narrow openings were incised in the horizontal PCB to form the vertical coils' intersections. The vertical coils were printed on small circuit boards and slotted into the incisions. The connections were fixed and assembled by soldering the lateral and horizontal coils. The vertical coils were fixed by glue. All loops were serially connected and fed by a single power supply. The fabricated structure consisted of an array of 3×3 elements in *x* and *y* directions and its effective powering area was 144 cm^2 (see Figure 5).

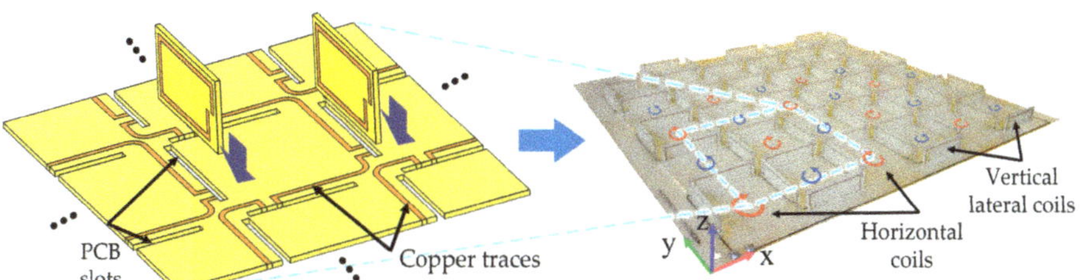

Figure 5. The vertical coils were slotted into the horizontal printed circuit board (PCB) openings. All the loops were serially connected. The connections were fixed by soldering and gluing.

Since the proposed structure is designed to work through near-field coupling between the Rx/Tx coils, the maximum frequency is limited such that the overall length of the Tx coil remains much lower than the wavelength. However, there is no minimum frequency bound as the coils can generate single-sided DC magnetic fields. A small multi-turn Rx coil was used as the magnetic probe to measure the mutual inductance under and on top of the Tx. As the spatial pattern of the single-sided Tx system is non-uniform, the mutual inductance is dependent on the Rx location, however, we considered the peak values. Table 1 presents the measured inductance of the involved coils.

Table 1. Measurement of the maximum inductance of the transmitter.

Object/Quantity	Rx Coil (Probe)	Tx Coil (3D Coils)
AC Resistance	$0.5\ \Omega$	$7.8\ \Omega$
Inductance	$11\ \mu H$	$9\ \mu H$
Mutual Ind. Peak at $z = +9.6$ mm	37.8 nH	
Mutual Ind. Peak at $z = -9.6$ mm	9.0 nH	

Wirelessly powered microrobots often include a multi-turn receptor coil whose impedance is modified by a tuning capacitance. To verify the single-sided performance of the introduced Tx setup, the spatial magnetic pattern of the proposed design was measured by a magnetic probe with a design similar to that of a typical power receptor coil in a swimming microrobot. The probe was connected to a spectrum analyzer (GW INSTEK GSP-810, New Taipei City, Taiwan) as well as an oscilloscope (RIGOL Ds4012, Beijing, China) to measure the field's intensity and polarity, respectively. A gridded polystyrene planar spacer was used to build different distance levels from the Tx. The Rx was positioned over the center of the gridded plane and manually scanned over the pad in 10 mm steps. The typical setup for a microrobot swimming above the Tx scheme and the magnetic field measurement setup are shown in Figure 6.

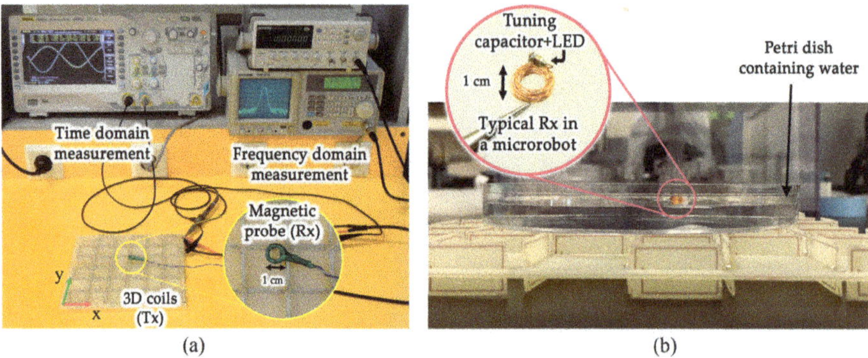

(a)   (b)

Figure 6. (**a**) The setup for measuring spatial patterns of the magnetic near-field at various distances from the transmitter (Tx) coil (gridded spacers not shown) at 1000 kHz. The magnetic probe had a similar structure to the microrobot's power receiver coil. (**b**) The typical placement of a microrobot's receiver (Rx) coil swimming on water. It is supplied by the single-sided power transmitter.

The absolute efficiency of WPT in the strengthened side highly depends on the design of on the Rx structure and may not yield information about shielding. However, to evaluate the shielding function, the magnetic field's intensity was measured at different distances from the coil array to calculate the relative WPT efficiency above and below the proposed 3D coil array.

According to the efficiency and shielding definition given by Paul [7], the WPT efficiency at different levels below the proposed Tx decreased 15-fold compared to the corresponding levels above the structure. With a shielding effectiveness of 11.7 dB for perpendicular magnetic fields, the proposed setup is a good solution for WPT to targets enclosed by radiation-sensitive electronic facilities. The measured radiation map (shown in Figure 7) had a dynamic range of 34 dB (from the pattern's null to peak) and is in very good agreement with the simulation results.

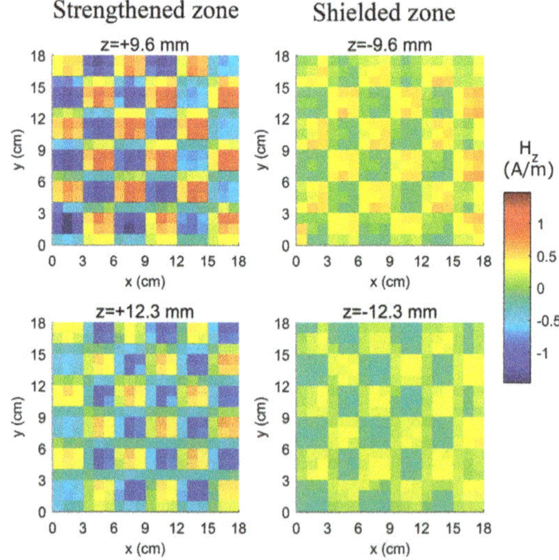

Figure 7. Magnetic radiation map at $z = \pm 9.6$ mm and ± 12.3 mm. The power transferred to the shorted probe on the top side is 15 times greater than that transferred to the bottom side.

The proposed structure is free of magnetizable elements such as ferrite shields and iron cores; hence, the actuating current can be maximized to the maximum tolerable temperature of the PCB substrate. Similar to other electromagnets, the current can be significantly increased by optimizing the trace width and thickness and by using cooling systems and particular substrates. As long as the overall length of the Tx coil is much lower than the wavelength of the working frequency, the effective powering area of this coil array can be extended in x and y directions.

4. Conclusions

Planar coils printed on circuit boards have been the major building block of most of wireless power transfer systems for many years. In this study, a new type of power transmitter coil has been developed by employing 3D current routes instead of planar printed coils. The coil scheme includes out-of-plane elements and is assembled based on a generalized Halbach array, which allows for magnetic streams to construct and cancel on different sides of the actuator without the use of any shielding material. The controlled radiation pattern of the proposed structure will find a wide range of applications, from wireless powering to biomedical facilities utilizing MRI-driven microrobots in environments that are sensitive to electromagnetic interference. The proposed structure eliminates the need for traditional magnetic shields and enables the miniaturization of the wireless charging pads. The measurements confirm that the power level emitted through the back lobe was 11.7 dB lower than that of the front lobe. In addition to the quantified radiation pattern, the powering function of the 3D coil was verified by a microrobot swimming above the structure. This paper validates a general approach to the design of a low leakage power transmitter, though the powering efficiency can still be enhanced with regard to the particular needs of various applications. Other structures besides cubic arrays in a chessboard-shaped combination, such as an isometric or honeycomb structure, may also be employed.

Author Contributions: A.H. and S.A. conceived and designed the experiments; A.H. performed the experiments; A.H. and S.A. analyzed the data; A.H. contributed materials/analysis tools; A.H. and S.A. wrote the paper.

Acknowledgments: This work was supported by the National Research Foundation of Korea (NRF) grant funded by the Korea government (MSIT)(2017R1A1A1A05001350) and National Research Foundation of Korea (NRF) grant funded by the Korean government (MSIP)(2017R1A5A1015596).

Conflicts of Interest: The authors declare no conflict of interest.

References

1. Shin, J.; Shin, S.; Kim, Y.; Ahn, S.; Lee, S.; Jung, G.; Jeon, S.-J.; Cho, D.H. Design and Implementation of Shaped Magnetic Resonance Based Wireless Power Transfer System for Roadway-Powered Moving Electric Vehicles. *IEEE Trans. Ind. Electron.* **2014**, *61*, 1179–1192. [CrossRef]
2. Kim, D.; Hwang, K.; Park, J.; Park, H.H.; Ahn, S. High-Efficiency Wireless Power and Force Transfer for a Micro-Robot Using a Multiaxis AC/DC Magnetic Coil. *IEEE Trans. Magn.* **2017**, *53*, 2015. [CrossRef]
3. Xue, R.F.; Cheng, K.W.; Je, M. High-efficiency wireless power transfer for biomedical implants by optimal resonant load transformation. *IEEE Trans. Circuits Syst. I Regul. Pap.* **2013**, *60*, 867–874. [CrossRef]
4. Ramrakhyani, K.; Mirabbasi, S.; Chiao, M. Design and optimization of resonance-based efficient wireless power delivery systems for biomedical implants. *IEEE Trans. Biomed. Circuits Syst.* **2011**, *5*, 48–63. [CrossRef]
5. Chen, C.-J.; Chu, T.-H.; Lin, C.-L.; Jou, Z.-C. A Study of Loosely Coupled Coils for Wireless Power Transfer. *IEEE Trans. Circuits Syst. II Express Briefs* **2010**, *57*, 536–540. [CrossRef]
6. Kiani, M.; Jow, U.M.; Ghovanloo, M. Design and Optimization of a 3-Coil Inductive Link for Efficient Wireless Power Transmission. *IEEE Trans. Biomed. Circuits Syst.* **2011**, *5*, 579–591. [CrossRef]
7. Paul, C.R. *Introduction to Electromagnetic Compatibility*; Wiley: Hoboken, NJ, USA, 1992.
8. Kim, H.; Song, C.; Kim, D.H.; Jung, D.H.; Kim, I.M.; Kim, Y.I.; Kim, J.; Ahn, S.; Kim, J. Coil design and measurements of automotive magnetic resonant wireless charging system for high-efficiency and low magnetic field leakage. *IEEE Trans. Microw. Theory Tech.* **2016**, *64*, 383–400. [CrossRef]
9. IEEE-SA Standards Board. *IEEE Standard for Safety Levels With Respect to Human Exposure to Radio Frequency Electromagnetic Fields, 3 kHz to 300 GHz*; The Institute of Electrical and Electronics Engineers, Inc.: New York, NY, USA, 2005.

10. Smoots, K.; Vogel, R.L. *Cardiac Pacing and Defibrillation in Paediatric and Congenital Heart Disease; First*; John Wiley & Sons: Hoboken, NJ, USA, 2017.

11. Hajiaghajani, A.; Hashemi, S.; Abdolali, A. Adaptable Setups for Magnetic Drug Targeting in Human Muscular Arteries: Design and Implementation. *J. Magn. Magn. Mater.* **2017**, *438C*, 173–180. [CrossRef]

12. Hajiaghajani, A.; Abdolali, A. Magnetic field pattern synthesis and its application in targeted drug delivery: Design and implementation. *Bioelectromagnetics* **2018**, *338*, 325–338. [CrossRef]

13. Hashemi, S.; Hajiaghajani, A.; Abdolali, A. Noninvasive Blockade of Action Potential by Electromagnetic Induction. Available online: https://arxiv.org/ftp/arxiv/papers/1809/1809.06199.pdf (accessed on 13 March 2019).

14. Kósa, G.; Jakab, P.; Székely, G.; Hata, N. MRI driven magnetic microswimmers. *Biomed. Microdevices* **2012**, *14*, 165–178. [CrossRef]

15. Kósa, G.; Jakab, P.; Jolesz, F.; Hata, N. Swimming Capsule Endoscope using static and RF magnetic field of MRI for propulsion. In Proceedings of the 2008 IEEE International Conference on Robotics and Automation, Pasadena, CA, USA, 19–23 May 2008; pp. 2922–2927.

16. Martel, S.; Felfoul, O.; Mathieu, J.-B.; Chanu, A.; Tamaz, S.; Mohammadi, M.; Mankiewicz, M.; Tabatabaei, N. MRI-based Medical Nanorobotic Platform for the Control of Magnetic Nanoparticles and Flagellated Bacteria for Target Interventions in Human Capillaries. *Int. J. Rob. Res.* **2009**, *28*, 1169–1182. [CrossRef]

17. Vartholomeos, P.; Fruchard, M.; Ferreira, A.; Mavroidis, C. MRI-guided nanorobotic systems for therapeutic and diagnostic applications. *Annu. Rev. Biomed. Eng.* **2011**, *13*, 157–184. [CrossRef]

18. Kim, D.; Hwang, K.; Park, J.; Park, H.H.; Ahn, S. Miniaturization of implantable micro-robot propulsion using a wireless power transfer system. *Micromachines* **2017**, *8*, 269. [CrossRef]

19. Uvet, H.; Demircali, A.A.; Kahraman, Y.; Varol, R.; Kose, T.; Erkan, K. Micro-UFO (untethered floating object): A highly accurate microrobot manipulation technique. *Micromachines* **2018**, *9*, 126. [CrossRef]

20. Fu, Q.; Zhang, S.; Guo, S.; Guo, J. Performance Evaluation of a Magnetically Actuated Capsule Microrobotic System for Medical Applications. *Micromachines* **2018**, *9*, 641. [CrossRef]

21. Bi, C.; Guix, M.; Johnson, B.V.; Jing, W.; Cappelleri, D.J. Design of microscale magnetic tumbling robots for locomotion in multiple environments and complex terrains. *Micromachines* **2018**, *9*, 68. [CrossRef]

22. Honda, T.; Arai, K.I.; Ishiyama, K. Micro swimming mechanisms propelled by external magnetic fields. *IEEE Trans. Magn.* **1996**, *32*, 5085–5087. [CrossRef]

23. Jawad, A.M.; Nordin, R.; Gharghan, S.K.; Jawad, H.M.; Ismail, M. Opportunities and challenges for near-field wireless power transfer: A review. *Energies* **2017**, *10*, 1022. [CrossRef]

24. Balanis, C.A. *Antenna Theory: Analysis and Design*; John Wiley & Sons: Hoboken, NJ, USA, 2012.

25. Kim, J.; Kim, J.; Kong, S.; Kim, H.; Suh, I.S.; Suh, N.P.; Cho, D.H.; Kim, J.; Ahn, S. Coil design and shielding methods for a magnetic resonant wireless power transfer system. *Proc. IEEE* **2013**, *101*, 1332–1342. [CrossRef]

26. Si, P.; Hu, P.; Malpas, S.; Budgett, D. A frequency control method for regulating wireless power to implantable devices. *Biomed. Circuits Syst.* **2008**, *2*, 22–29. [CrossRef]

27. Hajiaghajani, A.; Abdolali, A. Patterning of Subwavelength Magnetic Fields Along a Line by Means of Spatial Spectrum: Design and Implementation. *IEEE Magn. Lett.* **2017**, *8*, 1–4. [CrossRef]

28. Abdolali, A.; Mohtadi Jafari, A. Flexible Control of Magnetic Fields by Shaped-Optimized Three Dimensional Coil Arrays. *IEEE Magn. Lett.* **2018**, *10*, 1–5. [CrossRef]

29. Jafari, A.M.; Abdolali, A. Manipulation of the electromagnetic near-fields by 3D printed coils: from design to fabrication. *IET Microw. Antennas Propag.* **2018**, *12*, 1461–1465. [CrossRef]

30. Halbach, K. Application of permanent magnets in accelerators and electron storage rings. *J. Appl. Phys.* **1985**, *57*, 3605–3608. [CrossRef]

31. Barnsley, L.C.; Carugo, D.; Stride, E. Optimized shapes of magnetic arrays for drug targeting applications. *J. Phys. D Appl. Phys.* **2016**, *49*, 225501. [CrossRef]

32. Barnsley, L.C.; Carugo, D.; Owen, J.; Stride, E. Halbach arrays consisting of cubic elements optimised for high field gradients in magnetic drug targeting applications. *Phys. Med. Biol.* **2015**, *60*, 8303–8327. [CrossRef]

33. Nacev, A. *Magnetic Drug Targeting: Developing the Basics*; University of Maryland, College Park: College Park, MD, USA, 2013.

34. Kim, H.; Hwang, K.; Park, J.; Kim, D.; Ahn, S. Design of single-sided AC magnetic field generating coil for wireless power transfer. In Proceedings of the 2017 IEEE Wireless Power Transfer Conference (WPTC), Taipei, Taiwan, 10–12 May 2017; pp. 1–3.
35. Griffiths, D.J. *Introduction to Electrodynamics*; Prentice Hall: Upper Saddle River, NJ, USA, 1999.
36. Overfelt, P.L. Near fields of the constant current thin circular loop antenna of arbitrary radius. *IEEE Trans. Antennas Propag.* **1996**, *44*, 166–171. [CrossRef]

 *micromachines*

Article

A Miniaturized Circularly-Polarized Antenna for In-Body Wireless Communications

Yi Fan [1], Xiongying Liu [2,*], Jiming Li [2] and Tianhai Chang [2]

[1] School of Electronics and Information, Guangdong Polytechnic Normal University, Guangzhou 510665, China; hnanfy@163.com

[2] School of Electronic and Information Engineering, South China University of Technology, Guangzhou 510641, China; hnanlxy@163.com (J.L.); thchang@scut.edu.cn (T.C.)

* Correspondence: liuxy@scut.edu.cn; Tel.: +86-20-8711-1435

Received: 5 December 2018; Accepted: 16 January 2019; Published: 19 January 2019

Abstract: A novel miniaturized single-fed circularly-polarized (CP) microstrip patch antenna operating in the Industrial, Scientific, Medical (ISM) band of 2.40–2.48 GHz, is comprehensively proposed for implantable wireless communications. By employing reactive loading in the arrow-shaped slotted patch to form slow wave effect and embedding V-shaped slots into patch to lengthen the current path, the proposed implantable antenna is minimized with the overall dimensions of 9.2 mm × 9.2 mm × 1.27 mm. The radiation patterns of the proposed antenna illustrate the performance of left-handed circular polarization. The simulated results show that an impedance bandwidth of 7.2% (2.39–2.57 GHz) and an axial ratio bandwidth of 3.7% (2.39–2.48 GHz) at the ISM band are achieved, respectively. Ex vivo measured results are in good agreement with the corresponding simulated ones.

Keywords: circular polarization; implantable antenna; reactive loading; slow wave effect

---

1. Introduction

Implantable medical devices (IMDs) have increasingly caught the attention of the scientific community due to their wireless capabilities of detecting bio-medical information and transmitting health data much more flexibly and conveniently than traditional wired sensors placed exterior to the body [1–3]. These devices have been widely adopted in many applications including neural recording [4], glucose monitoring [5], and intracranial pressure monitoring [6], etc.

Implantable antennas act as a key factor to assure wireless communications between the implantable devices and the external equipment [7], and have been assigned at the industrial, scientific, and medical (ISM) bands with the operating frequency ranges of 433.1–434.8 MHz [8,9], 902–928 MHz [10], and 2.40–2.48 GHz [11,12]. Additionally, the 402–405 MHz medical implant communication services (MICS) band [13] and the 1395–1400 MHz wireless medical telemetry services (WMTS) band [14] are also designated for the implantable antennas. Implantable devices used for the biomedical telemetry are typical microsystem and therefore cannot accommodate large antennas. Due to the relatively short electromagnetic wavelength, the implantable antenna working at the ISM higher frequency band of 2.40–2.48 GHz is adopted by some researchers, making the dimension electrically small enough. Moreover, considerable design efforts, such as employing high-permittivity dielectric substrates [15], loading shorting pins to connect the patch and the ground [16], and extending the current flow path on the patch surface [17], have been made to realize the compact dimension of the implantable antennas. Many crucial factors such as biocompatibility, specific absorption rate (SAR), far-field radiation, and operating bandwidth, should be also considered during the designing of implantable antennas.

The planar inverted-F antenna (PIFA) was demonstrated as a useful prototype in the design of implantable antennas because of its structural simplicity and low profile [18]. The monopole antenna with the advantages of omnidirectional radiation pattern and wide bandwidth has been integrated with the implantable system [19]. In [20], a differentially-fed dual-band flexible antenna was proposed for ingestible capsule system. Nevertheless, the above-mentioned antennas are linearly polarized and dependent on the relative orientations between the transmitters and the external receivers.

A circularly-polarized (CP) antenna is preferred for the implantable devices because it can reduce multipath distortion and provide flexible mobility, compared with a linearly polarized antenna. However, only a few works focus on the miniaturized CP implantable antennas. Capacitive loadings were introduced in a square patch with a central squared slot to realize circular polarization for an implantable antenna [21]. An axial-mode multilayer helical circularly-polarized implantable antenna for ingestible capsule endoscope system was presented in [22]. Additionally, by cutting rectangular slots into the patch and adding open stubs in the annular ring, a CP bio-friendly implantable annular ring antenna was realized in [23].

A miniaturized CP squared patch antenna for implantable devices at the ISM band of 2.40–2.48 GHz was preliminarily introduced in [24]. After that, future works have been carried out. In this article, an in-depth study on the working principle and a more detailed analysis on the performance of the proposed antenna are presented while a proof-of-concept fabricated prototype was fully characterized, verifying the good performance of the proposed antenna topology. Through introducing reactive loading, slow wave effect is formed on the radiator, making the dimension of the proposed antenna compact.

The contents of this article are organized as follows: In Section 2, a simulation environment is set up, the geometry of the proposed implantable antenna is described, and the simulation results are studied. In Section 3, the miniaturization of working mechanism, CP properties, and parameter analysis are given. Section 4 presents the measured results before a useful conclusion is made in Section 5.

2. Antenna Design and Simulation

2.1. Simulation Environment

As shown in Figure 1, the proposed antenna is simulated in a three-layer tissue numerical model with the dimensions of 50 mm × 50 mm × 58 mm that imitates the real human environment. The human tissue is composed of skin, fat, and muscle. The electrical properties of the tissues vary with frequency. Table 1 lists the values of dielectric properties for skin, fat, and muscle in the simulated model at 2.45 GHz. To be closer to external devices and reduce the path loss in the tissue, the proposed antenna is implanted in a depth of 2 mm from the top of skin. The simulated tool employs ANSYS High Frequency Structure Simulator (HFSS) software (v.13, Ansys Inc., Canonsburg, PA, USA) for modeling, optimizing, and analyzing.

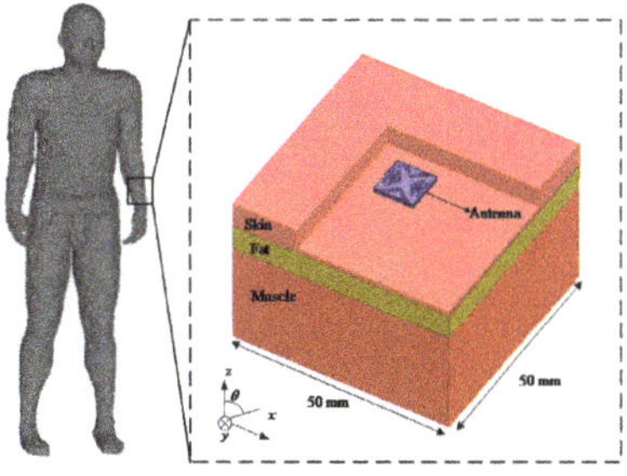

Figure 1. Simulation environment of the proposed antenna.

Table 1. Dielectric properties of different tissues at 2.45 GHz.

Tissues	Thickness (mm)	ε_r	σ (S/m)
Skin	4	38.0	1.46
Fat	4	5.28	0.1
Muscle	50	52.7	1.74

2.2. Geometry of the Proposed Circularly-Polarized Antenna

The configuration of the proposed implantable antenna is demonstrated in Figure 2, the dimensions of the patch are fixed to 9.0 mm × 9.0 mm with a ground plane of 9.2 mm × 9.2 mm. To achieve the miniaturization, the antenna is manufactured on a Rogers 3010 substrate with a high dielectric constant of ε_r = 10.2 and a low loss tangent of tan δ = 0.0035, covered by a layer of superstrate with the same material as the substrate, each with a thickness of H = 0.635 mm. The superstrate is utilized to separate human tissues from the conducting patch of the proposed antenna and to enhance the matching with the around inner tissues. In order to avoid shorting and to relieve mismatching, the proposed antenna should be wrapped by a thin film of biocompatible materials alumina (ε_r = 9.2, tan δ = 0.008). The 50-Ω coaxial cable feeding point is welded at the position (d, d) along the center of diagonal of the patch. Two small triangle patches are embedded on the upper and lower sides of the proposed antenna and connected with the main patch through two high impedance lines. Additionally, V-shaped slots are embedded into the left and right side of the proposed antenna. It should be noted that perturbation slots are cut to optimize the axial ratio (AR). Table 2 lists the detailed values of the geometrical dimension after the optimization with the aid of ANSYS HFSS.

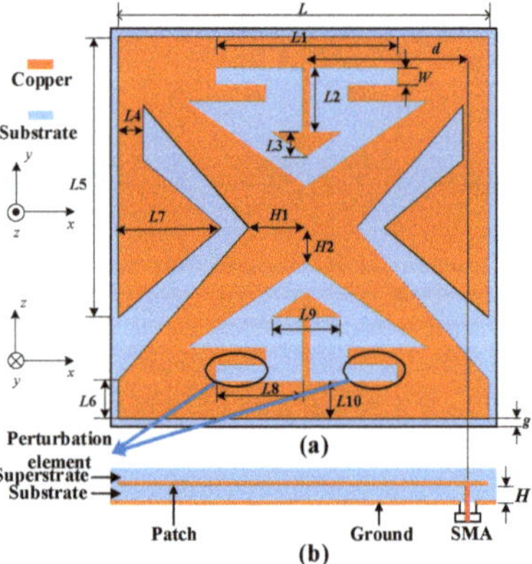

Figure 2. Geometry of the proposed antenna at: (**a**) Top view; and (**b**) side view.

Table 2. Dimensions of the proposed antenna parameters (unit: mm).

Symbol	Value	Symbol	Value	Symbol	Value
L	9.0	W	0.5	$L1$	4.6
$L2$	1.8	$L3$	0.6	$L4$	0.4
$L5$	6.4	$L6$	0.8	$L7$	2.6
$L8$	2.2	$L9$	0.9	d	3.5
g	0.1	$H1$	1.8	$H2$	1

2.3. Simulated Results

Figure 3 illustrates the simulated reflection coefficient together with the axial ratio (AR) of the proposed antenna in main radiation direction towards the outside of human body. The simulated impedance bandwidth is covered from 2.39 GHz to 2.57 GHz with S11 less than −10 dB while the AR bandwidth can be extended from 2.39 GHz to 2.48 GHz with AR below 3 dB.

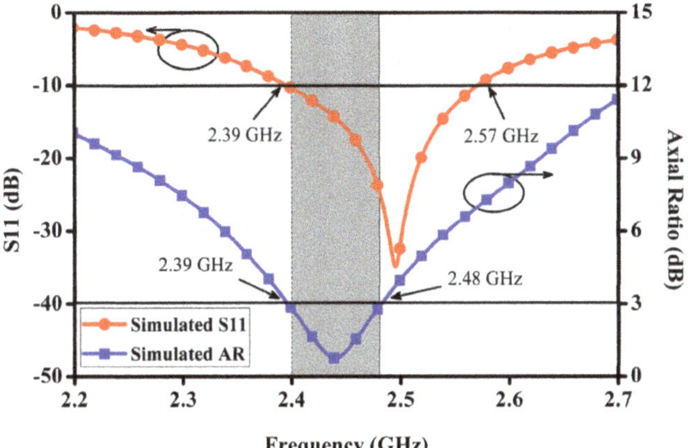

Figure 3. Simulated reflection coefficient and axial ratio (AR) varying with frequency.

Figure 4 depicts the simulated far-field gain radiation patterns of the proposed antenna in two principal planes (i.e., *xy*-plane and *xz*-plane) at 2.45 GHz. Its maximum left-handed circular polarization (LHCP) radiation is towards the antenna's boresight at theta = $0°$, that is the off-body direction as desired. Due to the fact that the proposed antenna is very compact and implanted in the lossy tissue with the limited space, different from the conventional antenna operating in free space, the peak realized gain is −24.8 dBi at 2.45 GHz.

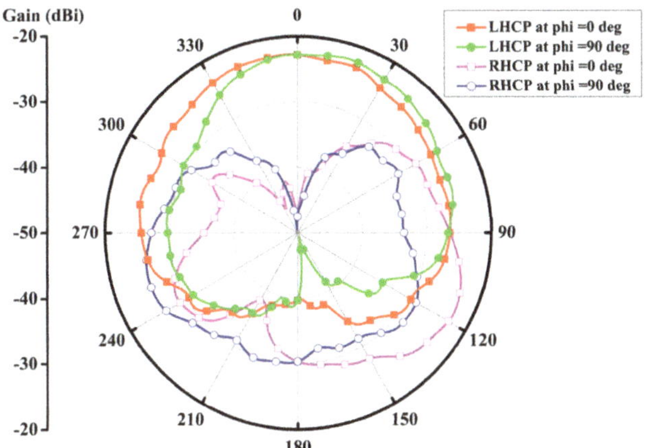

Figure 4. Simulated radiation patterns at 2.45 GHz.

3. Antenna Analysis

3.1. Miniaturization of the Proposed Antenna

In order to investigate the mechanism of miniaturization, the corresponding topology of the antenna is evolved from Case 1 to Case 4 by subsequently cutting slots and loading patch into a conventional square patch antenna, as shown in Figure 5. Here, all antennas keep the fixed sizes of 9.2 mm × 9.2 mm and the simulating settings are similar.

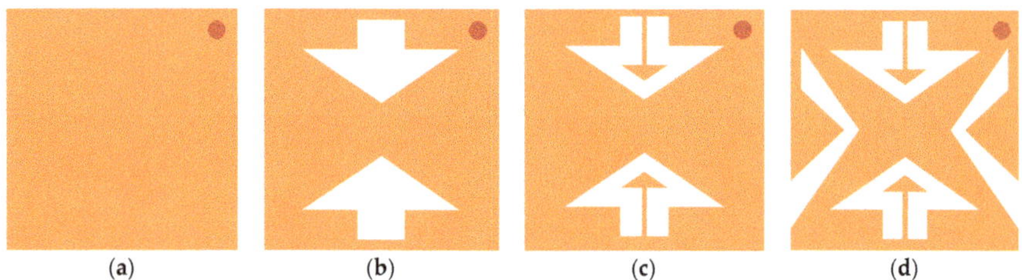

Figure 5. Evolving structures of the proposed antenna in: (**a**) Case 1; (**b**) Case 2; (**c**) Case 3; and (**d**) Case 4.

With reference to Figure 5, Case 1 is a conventional square patch antenna with the feeding port at the upper right diagonal, Case 2 is obtained by etching an arrow-shaped slot in the upper and lower parts of the patch. In Case 3, two small triangle patches are respectively loaded on the arrow-shaped slot of Case 2 and connected with the main patch through two high impedance lines. As shown in Figure 6, the resonant frequency of 4 GHz in Case 1 shifts down to 3.6 GHz in Case 2, then converts to 3.16 GHz in Case 3, indicating that a 21% of miniaturization can be achieved.

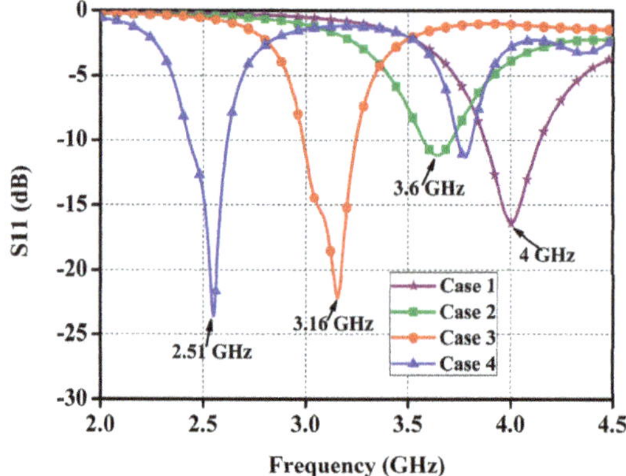

Figure 6. Simulated S11 of four cases embedded in the same phantom.

Theoretically, an antenna can be equivalent to a transmission line, as shown in Figure 7, which is characterized by a series inductance L_0 and a shunt capacitance C_0 per unit length. When the antenna is loaded with patch through a high impedance lines, the miniaturization of the proposed antenna can be achieved by taking advantage of the principle of slow waves [25]. The mechanism can be understood through calculating the propagation velocity as:

$$v_p = \frac{1}{\sqrt{L_0 C_0}} = \frac{c}{\sqrt{\varepsilon_{eff}}} = \lambda_g f \tag{1}$$

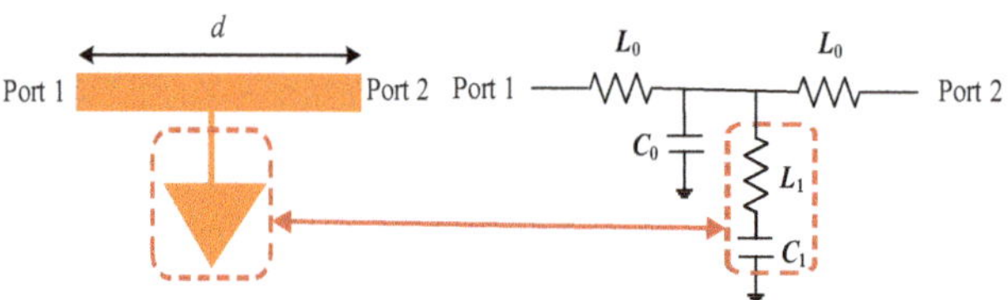

Figure 7. Transmission line model with LC loadings.

According to Equation (1), by adding triangle patches (equivalent to a capacitance C_1) and high impedance line (equivalent to an inductance L_1), total equivalent capacitance and/or inductance are increased, subsequently propagation velocity becomes slower, resulting in waveguide wavelength smaller when the frequency remains unchanged.

To further miniaturize the proposed antenna, V-shaped patches are etched into the left and right sides of Case 3, lengthening the current path, as established in Case 4 of Figure 5d. Compared with Case 1, the corresponding resonate frequency of Case 4 is shifted down from 4 GHz to 2.51 GHz, demonstrating that 37.3% of miniaturization is achieved.

3.2. CP property of the Antenna

As shown in Figure 2, perturbation slots are introduced in the patch to strengthen the CP performance. With reference to Figure 8, it can be seen that the small slots take critical role in the impedance matching and circular polarization. For the purpose of visualizing how the circular polarization is generated, the simulated surface current distributions on the patch at 2.45 GHz for four moments of 0T, T/4, T/2, and 3/4T are demonstrated in Figure 9. With the increment of time by a step of T/4, the currents rotate clockwise, transmitting LHCP waves in the boresight direction.

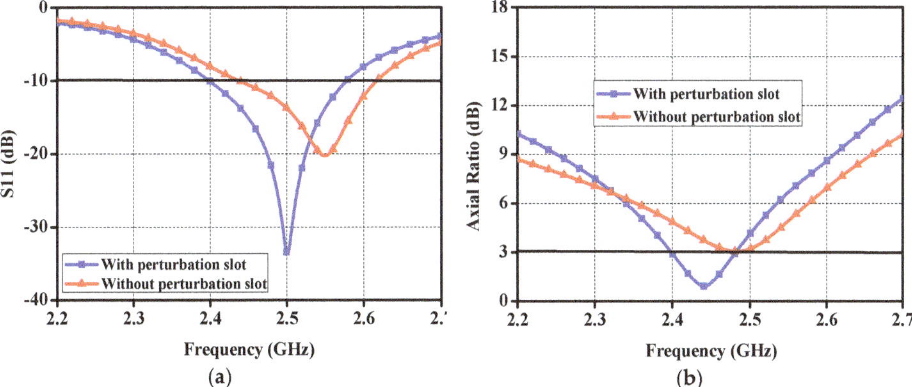

Figure 8. Performance of the proposed antenna with/without perturbation slots: (**a**) Reflection coefficient; and (**b**) AR.

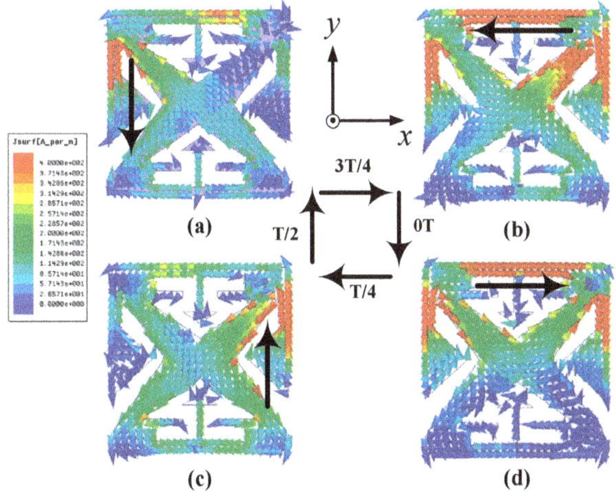

Figure 9. Current distributions on the patch at: (**a**) t = 0T; (**b**) t = T/4; (**c**) t = T/2; and (**d**) t = 3T/4.

3.3. Parameter Studies

To obtain available guidelines for the practical design of the proposed antenna, various important parameters that can influence the return loss and axial ratio at the boresight direction are examined. As a key parameter is studied, the other parameters are kept constant. The width *L4* of the high impedance microstrip line shows the crucial influence on the reflection coefficient and AR, due to its effect on the current distributions on the patch of the proposed antenna. As exhibited in Figure 10, the return loss and axial ratio are sensitive to different *L4* values. A reasonable axial ratio at 2.45 GHz can be achieved when *L4* is set to be 0.4 mm. If *L4* becomes larger or smaller, there is a drastic influence on the AR. Figure 11 demonstrates the effect of tuning *L5* on the performance of the proposed antenna. With reference to the curves in the figure, a linear increase in *L5* (from 6.3 mm to 6.5 mm) will result in shifting the resonance down to the lower frequency and deviating the CP towards the lower band.

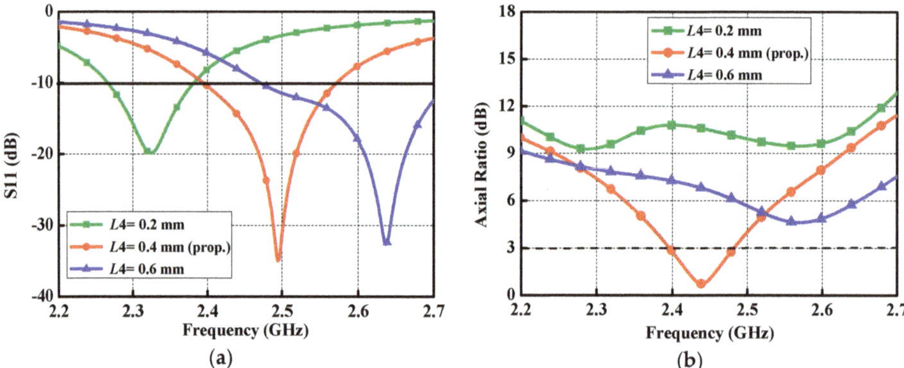

Figure 10. Performance of the proposed antenna with different *L4* values: (**a**) Reflection coefficient; and (**b**) AR.

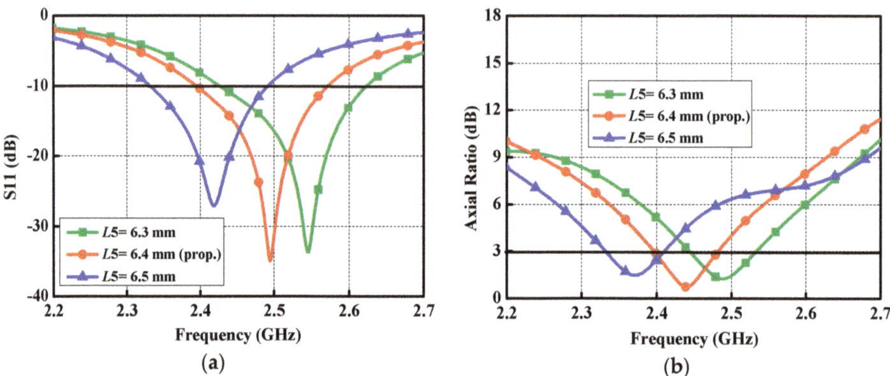

Figure 11. Performance of the proposed antenna with different *L5* values: (**a**) Reflection coefficient; and (**b**) AR.

3.4. Safety Consideration

When the proposed antenna is implanted into the human body, specific absorption rate (SAR) for safety concerns should be evaluated. The IEEE C95.1-2005 standard limits the SAR average over any 10 g of tissue in the shape of a cube to less than 2 W/Kg (SAR 10g, max $\leq$ 2 W/kg) [26]. Therefore, through calculation, the maximum 10-g averaged SAR value is 81.5 W/kg at 2.45 GHz on condition that the power delivered to the proposed antenna is set to be 1 W, meaning that the delivered power should be below 24.5 mW to meet the IEEE C95.1-1999 standard.

4. Experimental Results

In order to validate the design strategy, the prototype of the proposed implanted antenna was fabricated and assembled. The measurement environment surrounding the proposed antenna is a piece of fresh streaky pork (shortly after slaughter) comprised layers of skin, fat, and muscle. Figure 12 shows the photos of fabricated antenna and measurement setup. The S-parameters of the antenna against frequency are measured with the aid of an Agilent N5230A vector network analyzer (VNA) (Keysight Technologies, Santa Rosa, CA, USA). The simulated and measured S-parameters are shown in Figure 13. The measured bandwidth for S11 < −10 dB is 12.9%, covering 2.32 to 2.64 GHz. There is a slight difference between the simulated and measured results mainly due to the tolerances in the fabrication process and measurement. A linearly polarized dipole, working as a receiver, is placed 150 mm away from the proposed antenna to testify the CP property of the proposed antenna. The S21 between the proposed antenna and the dipole was measured when the dipole was placed at the phi = −45°, 0°, 45°, and 90°, respectively. With reference to Figure 13, there is a maximum deviation only up to 3 dB for the measured S21, proving the CP property of the proposed antenna.

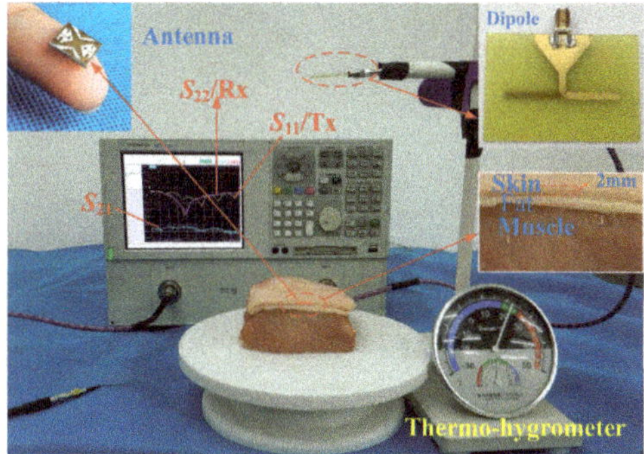

Figure 12. Photograph of the fabricated antenna prototype and measurement setup.

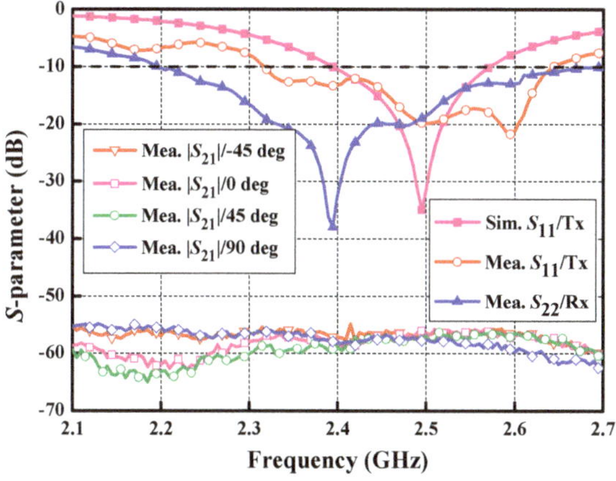

Figure 13. Measured S-parameters of the proposed antenna.

5. Conclusions

This article has numerically designed and experimentally studied a novel miniaturized single-fed circularly-polarized microstrip patch antenna at the ISM band of 2.40–2.48 GHz. By introducing reactive loading to form slow wave effect on the radiator and etching V-shaped slots on the main patch to lengthen the current flow path, a miniaturized antenna with the dimensions of 9.2 mm × 9.2 mm × 1.27 mm can be obtained. The radiations of the proposed antenna show a left-handed circular polarization when we adjust the sizes of geometrical structure. The prototype of the proposed implantable antenna has been implemented. The agreements between the simulated results and ex vivo measured ones have been reached.

As illustrated in Table 3, the performance of the proposed antenna is not perfect. Whereas, a trade-off has been reached. Compared with [5], although the dimensions of the proposed antenna are bigger, its AR bandwidth is broadened. The proposed antenna has less efficiency than the work in [21], but it has smaller dimensions and bigger AR bandwidth; compared with the art [23], the proposed antenna has compact dimensions with higher AR bandwidth; compared with [27], the proposed antenna has much compact dimensions with high efficiency. Hence, based on the slow wave effect, this novel implantable CP antenna is designed and miniaturized with the advantages of great size reduction and high polarization purity. The merit performance of the proposed implantable antenna shows the great potential in the application of biomedical telemetry, such as subcutaneous real-time glucose monitoring.

Table 3. Comparison of the proposed antenna with prior art.

Ref.	Dimensions (mm × mm × mm)	Bandwidth (S11 < −10 dB)	Bandwidth (AR < 3 dB)	Peak Gain (dBi)
[5]	$8.5 \times 8.5 \times 1.27$ (92 mm^3)	2.32–2.62 GHz (~12.2%)	2.42–2.48 GHz (~2.4%)	−17
[21]	$10 \times 10 \times 1.27$ (127 mm^3)	2.36–2.55 GHz (~7.7%)	2.44–2.48 GHz (~1.6%)	−22
[23]	$\pi \times (5.5)^2 \times 1.27$ (~120 mm^3)	2.31–2.51 GHz (~8.3%)	2.42–2.48 GHz (~2.49%)	−22.7
[27]	$10 \times 10 \times 1.27$ (127 mm^3)	2.35–2.50 GHz (~6.2%)	2.36–2.56 GHz (~8.13%)	−27.2
This work	$9.2 \times 9.2 \times 1.27$ (107 mm^3)	2.39–2.57 GHz (~7.2%)	2.39–2.48 GHz (~3.7%)	−24.8

Author Contributions: X.L. and Y.F. conceived the initial project; J.L. performed the design and experiment; Y.F. verified the results; Y.F. and J.L. prepared the original draft; Y.F. answered the comments and modified the manuscript; T.C. reviewed the manuscript; X.L. supervised the project; all authors read and approved the final manuscript.

Funding: This work was funded by the National Natural Science Foundation of China (No. 61372008), the Science and Technology Planning Project of Guangdong Province (Nos. 2014A010103014 and 2015B010101006), and the China Scholarship Council (No. 201706155018).

Acknowledgments: The authors would also like to thank the anonymous reviewers for their valuable comments and suggestions in improving the quality of this paper.

Conflicts of Interest: The authors declare no conflict of interest.

References

1. Kiourti, A.; Nikita, K.S. A review of implantable patch antennas for biomedical telemetry: Challenges and solutions. *IEEE Trans. Antennas Propag. Mag.* **2012**, *54*, 210–228. [CrossRef]
2. Soontornpipit, P.; Furse, C.M.; Chung, Y.C. Design of implantable microstrip antenna for communication with medical implants. *IEEE Trans. Microw. Theory Tech.* **2004**, *52*, 1944–1951. [CrossRef]
3. Damis, H.A.; Khalid, N.; Mirzavand, R.; Chung, H.; Mousavi, P. Investigation of epidermal loop antennas for biotelemetry IoT applications. *IEEE Access.* **2018**, *6*, 15806–15815. [CrossRef]
4. Neihart, N.M.; Harrison, R.R. Micropower circuits for bidirectional wireless telemetry in neural recording applications. *IEEE Trans. Biomed. Eng.* **2005**, *52*, 1950–1959. [CrossRef] [PubMed]

5. Liu, X.Y.; Wu, Z.T.; Fan, Y.; Tentzeris, E.M. A miniaturized CSRR loaded wide-beamwidth circularly polarized implantable antenna for subcutaneous real-time glucose monitoring. *IEEE Antennas Wirel. Propag. Lett.* **2017**, *16*, 577–580. [CrossRef]

6. Shah, S.A.A.; Yoo, H. Scalp-implantable antenna systems for intracranial pressure monitoring. *IEEE Trans. Antennas Propag.* **2018**, *66*, 2170–2173. [CrossRef]

7. Hall, P.S.; Hao, Y. *Antennas and Propagation for Body-Centric Wireless Communications*, 2nd ed.; Artech House: Norwood, MA, USA, 2012.

8. Huang, F.J.; Lee, C.M.; Chang, C.L.; Chen, L.K.; Yo, T.C.; Luo, C.H. Rectenna application of miniaturized implantable antenna design for triple-band biotelemetry communication. *IEEE Trans. Antennas Propag.* **2011**, *59*, 2646–2653. [CrossRef]

9. Duan, Z.; Guo, Y.X.; Xue, R.F.; Je, M.; Kwong, D.L. Differentially-fed dual-band implantable antenna for biomedical applications. *IEEE Trans. Antennas Propag.* **2012**, *60*, 5587–5595. [CrossRef]

10. Kiourti, A.; Nikita, K.S. Miniature scalp-implantable antennas for telemetry in the MICS and ISM bands: Design, safety considerations and link budget analysis. *IEEE Trans. Antennas Propag.* **2012**, *60*, 3568–3575. [CrossRef]

11. Merli, F.; Bolomey, L.; Zurcher, J.F.; Corradini, G.; Meurville, E.; Skrivervik, A.K. Design, realization and measurements of a miniature antenna for implantable wireless communication systems. *IEEE Trans. Antennas Propag.* **2011**, *59*, 3544–3555.

12. Warty, R.; Tofighi, M.R.; Kawoos, U.; Rosen, A. Characterization of implantable antennas for intracranial pressure monitoring: Reflection by and transmission through a scalp phantom. *IEEE Trans. Microw. Theory Tech.* **2008**, *56*, 2366–2376. [CrossRef]

13. Xu, L.J.; Guo, Y.X.; Wu, W. Miniaturized dual-band antenna for implantable wireless communications. *IEEE Antennas Wirel. Propag. Lett.* **2014**, *13*, 1160–1163.

14. Izdebski, P.M.; Rajagopalan, H.; Rahmat-Samii, Y. Conformal ingestible capsule antenna: A novel chandelier meandered design. *IEEE Trans. Antennas Propag.* **2009**, *57*, 900–909. [CrossRef]

15. Liu, Y.; Chen, Y.; Lin, H.; Juwono, F.H. A novel differentially fed compact dual-band implantable antenna for biotelemetry applications. *IEEE Antennas Wirel. Propag. Lett.* **2016**, *15*, 1791–1794. [CrossRef]

16. Xu, L.J.; Guo, Y.X.; Wu, W. Dual-band implantable antenna with open-end slots on ground. *IEEE Antennas Wirel. Propag. Lett.* **2012**, *11*, 1564–1567. [CrossRef]

17. Liu, C.; Guo, Y.X.; Xiao, S. Compact dual-band antenna for implantable devices. *IEEE Antennas Wirel. Propag. Lett.* **2012**, *11*, 1508–1511.

18. Liu, W.C.; Chen, S.H.; Wu, C.M. Bandwidth enhancement and size reduction of an implantable PIFA antenna for biotelemetry devices. *Microw. Opt. Technol. Lett.* **2009**, *51*, 755–757. [CrossRef]

19. Jung, Y.H.; Qiu, Y.; Lee, S.B.; Shih, T.Y.; Xu, Y.; Xu, R.; Lee, J.; Schendel, A.A.; Lin, W.; Williams, J.C.; et al. A compact parylene-coated WLAN flexible antenna for implantable electronics. *IEEE Antennas Wirel. Propag. Lett.* **2016**, *15*, 1382–1385. [CrossRef]

20. Duan, Z.; Guo, Y.X.; Je, M.; Kwong, D.L. Design and in vitro test of a differentially fed dual-band implantable antenna operating at MICS and ISM bands. *IEEE Trans. Antennas Propag.* **2014**, *62*, 2430–2439. [CrossRef]

21. Liu, C.R.; Guo, Y.X.; Xiao, S.Q. Capacitively loaded circularly polarized implantable patch antenna for ISM band biomedical applications. *IEEE Trans. Antennas Propag.* **2014**, *62*, 2407–2417. [CrossRef]

22. Liu, C.; Guo, Y.X.; Xiao, S.Q. Circularly polarized helical antenna for ISM-band ingestible capsule endoscope systems. *IEEE Trans. Antennas Propag.* **2014**, *62*, 6027–6039. [CrossRef]

23. Li, R.; Guo, Y.X.; Zhang, B.; Du, G.H. A miniaturized circularly polarized implantable annular-ring antenna. *IEEE Antennas Wirel. Propag. Lett.* **2017**, *16*, 2566–2569. [CrossRef]

24. Li, J.M.; Chang, T.H.; Liu, X.Y. A compact circularly polarized antenna for in-body wireless communications. In Proceedings of the 2017 IEEE International Symposium on Antennas and Propagation & USNC/URSI National Radio Science Meeting, San Diego, CA, USA, 9–14 July 2017; pp. 2593–2594.

25. Chi, P.; Waterhouse, R.; Itoh, T. Antenna miniaturization using slow wave enhancement factor from loaded transmission line models. *IEEE Trans. Antennas Propag.* **2011**, *59*, 48–57. [CrossRef]

26. *IEEE Standard for Safety Levels with Respect to Human Exposure to Radio Frequency Electromagnetic Fields, 3 kHz to 300 GHz, IEEE Standard C95.1-2005*; IEEE: New York, NY, USA, 2006; pp. 1–238.

27. Yang, Z.J.; Xiao, S.Q.; Zhu, L.; Wang, B.Z.; Tu, H.L. A Circularly polarized implantable antenna for 2.4-GHz ISM band biomedical applications. *IEEE Antennas Wirel. Propag. Lett.* **2017**, *16*, 2554–2557. [CrossRef]

Article

A Compact Broadband Antenna with Dual-Resonance for Implantable Devices

Rongqiang Li, Bo Li, Guohong Du, Xiaofeng Sun and Haoran Sun *

College of Electronic Engineering, Chengdu University of Information Technology, Chengdu 610225, China; liyq2011@cuit.edu.cn (R.L.); throbb@163.com (B.L.); dghong@cuit.edu.cn (G.D.); sunxf@cuit.edu.cn (X.S.)
* Correspondence: sunhaoran@cuit.edu.cn; Tel.: +86-28-8596-6349

Received: 12 November 2018; Accepted: 14 January 2019; Published: 16 January 2019

Abstract: A compact broadband implantable patch antenna is designed for the field of biotelemetry and experimentally demonstrated using the Medical Device Radiocommunications Service (MedRadio) band (401–406 MHz). The proposed antenna can obtain a broad impedance bandwidth by exciting dual-resonant frequencies, and has a compact structure using bent metal radiating strips and a short strategy. The total volume of the proposed antenna, including substrate and superstrate, is about 479 mm^3 (23 × 16.4 × 1.27 mm^3). The measured bandwidth is 52 MHz (382–434 MHz) at a return loss of −10 dB. The resonance, radiation and specific absorption rate (SAR) performance of the antenna are examined and characterized.

Keywords: implantable antenna; patch antenna; biotelemetry; specific absorption rate (SAR); Medical Device Radiocommunications Service (MedRadio) band

---

1. Introduction

Biotelemetry provides transmission of physiological signals from inside the human body to outside it, or vice versa. One of its applications is in the field of implantable medical devices (IMDs) [1]. An implantable antenna is a key component to communicate wirelessly between external equipment and IMDs in a human body [2,3]. Usually, an implantable antenna is used in the medical implant communications service (MICS) band (402–405 MHz), which was extended to the Medical Device Radiocommunications Service (MedRadio) band (401–406 MHz) in 2009. The implantable antenna in the human body faces many challenges, such as miniaturization, bandwidth, biocompatibility, and human body safety considerations.

Planar inverted-F antennas (PIFAs) have been used in the design of implantable antennas by many research groups because of their advantages [4–9]. Compared with conventional microstrip antennas, PIFAs have some desirable features, such as smaller size and nearly omnidirectional far-field radiating patterns. On the one hand, miniaturization is a key requirement for an implantable antenna due to some strict size constraints inside the human body. Kimi et al. [4] proposed two compact implanted antennas by using meandered and spiral structures, but they both have a larger size. Liu et al. [5] realized a miniaturized implantable antenna by embedding some slots in the ground plane. However, the slotted ground designs may be limited to application scenarios where there are no conductors on the ground plane. Li et al. [6] proposed a miniaturized implantable antenna by continually meandering two radiating edges of a square ring. We proposed a compact semi-circular implantable antenna by cutting three arc-shaped slots in a semi-circular patch [7]. However, the antennas in [6,7] have narrow bandwidth. On the other hand, an implantable antenna needs a wide impedance bandwidth to avoid frequency shift due to intersubject variations of the electrical properties of human body tissues and inaccuracies of fabrication and testing. To broaden the antenna impedance bandwidth, some methods have been used. Liu et al. [8] designed an implantable antenna by arranging a rectangular three-layer slotted patch structure. However, the stacked structure usually has a large thickness. A pi-shaped PIFA structure with two meandered strips was proposed for implantable biotelemetry in [9], but this antenna has a relatively large size. In addition to planar inverted F antennas, some other technologies have also been used to design wideband implantable antennas, such as monopole antennas [10], dipole antennas [11], loop antennas [12] and slot antennas [13]. An implantable broadband antenna was designed by combining a sigma-shaped monopole radiator and a novel C-shaped antenna to excite two modes [10]. Xu et al. [11] realized the bandwidth enhancement of a conformal implantable antenna by connecting a strip and a simple dipole to obtain dual-resonance. In [12], a flexible loop antenna by introducing three complementary split ring resonators (CSRRs) was used to improve the antenna impedance matching and obtain broad bandwidth. In [13], a coplanar waveguide-fed wideband dual-ring slot antenna was proposed to improve the antenna gain. Antennas in references [11–13] were designed as flexible and conformal structures for some special application scenarios.

In this paper, we propose a compact broadband implantable antenna with dual resonance for implantable medical devices. The metal radiating strips of the antenna were properly bent to obtain dual-resonant frequencies and a compact structure. Details of the antenna design and experimental results are presented and discussed.

2. Antenna Structure and Design

2.1. Antenna Structure

Figure 1 illustrates the geometry of the radiating metal layer, which consists of two mutually inverted rectangular opening rings and a C-shaped strip. The C-shaped strip has a short via and is surrounded by the inner of two opening rings. Thus, the proposed antenna can form a PIFA structure and resonate at a relatively low frequency. The antenna is fed by a 50 Ω coaxial cable, which is located at the internal rectangular opening ring. Figure 2 shows the one-layer skin simulation model based on [14] for the proposed implantable antenna, placed 3 mm from the top and bottom of the skin surface, and 40 mm from the other surfaces of skin. The electrical property parameters of this skin mode at 402 MHz are $\varepsilon_\gamma = 46.7$, $\sigma = 0.69$ s/m, in the simulation. This antenna is fabricated on a 0.635-mm-thick Rogers 6010 substrate, whose dielectric constant and loss tangent are respectively 10.2 and 0.0023, and covered by a superstrate of the same material. It is worth noting that, according to [14], the model in Figure 2 is a simplification of a three-layer model with similar return loss performance. In the three-layer model, the antenna is located under the skin adjacent to the fat layer. In reality, an antenna does not directly touch human tissue, which is placed in an implantable device, and the implantable device is placed in the human body. Some optimum parameters of the antenna derived using the high frequency structure simulator (HFSS) are given in Figure 1, and the other parameters are: $L_s = 3.6$ mm, $L_f = 15.8$ mm, $L_1 = 4.2$ mm, $L_2 = 10.7$ mm.

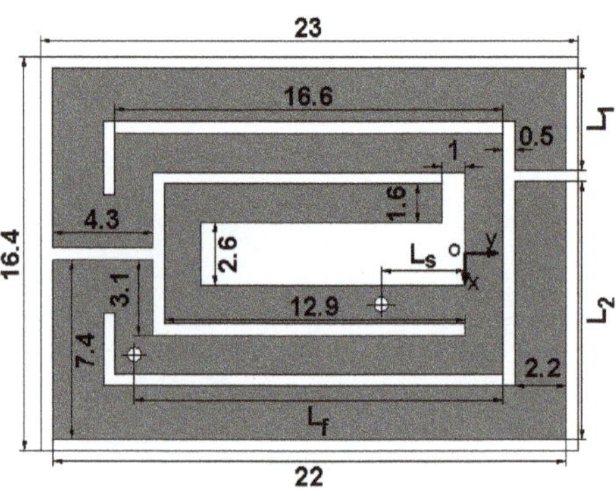

Figure 1. Geometry of the proposed implantable antenna (unit: mm).

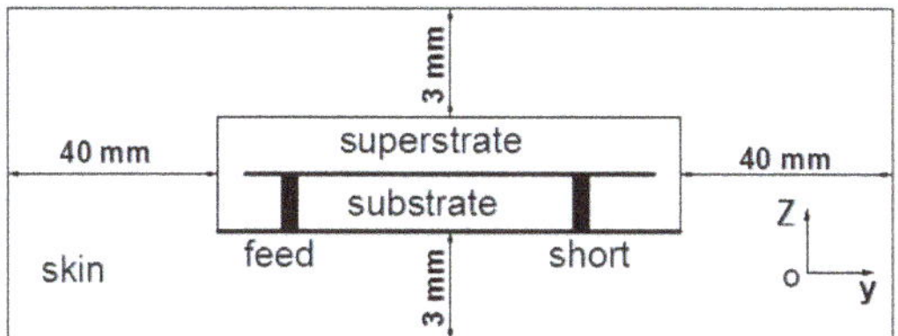

Figure 2. Side view of simulation model for the proposed implantable antenna.

2.2. Current Distributions of the Radiating Strips

In this work, the broad impedance bandwidth is due to a combination of two neighboring resonant frequencies at 391 MHz and 412 MHz. To better understand the operating mechanism, the current distributions of the proposed antenna at two resonant frequencies are shown in Figure 3. At 391 MHz, the current path of the inverted rectangular opening rings on the left side of the short via is opposite to one on the right side, and indicates that either part of the metal radiating strip resonates at this low frequency. At 412 MHz, the current of the two inverted rectangular opening rings flows in the same direction from one end to the other, and indicates that the whole metal strip contributes to the high resonant frequency. Thus, the antenna can realize dual resonances to obtain a broad impedance bandwidth. It is also worth mentioning that the current distribution at 402 MHz is similar to that at 412 MHz, which is not shown here for brevity.

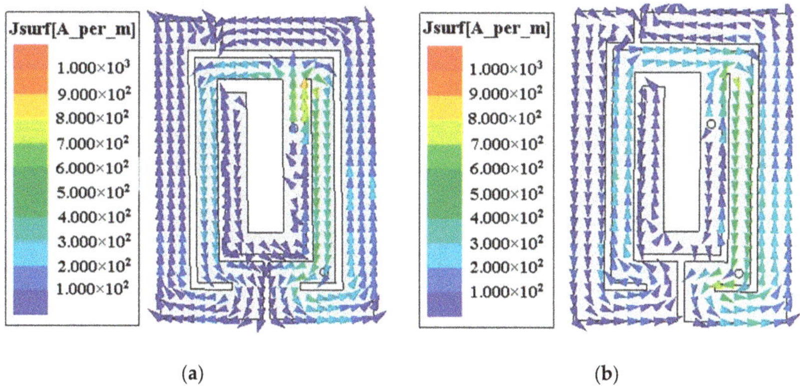

(a)   (b)

Figure 3. Current distributions of the proposed antenna at (**a**) 391 MHz; (**b**) 412 MHz.

2.3. Parametric Study

In order to achieve the proper resonant frequencies in Figure 1, parameters L_s (the short position), L_f (the feed position), and L_1 and L_2 (the tuned strip length) are analyzed by the electromagnetic simulator Ansoft HFSS. To accurately assess the influence of these parameters, only one parameter at a time is varied while others are kept constant. Figure 4a shows that the shorter the length L_s, the higher the low resonant frequency around 391 MHz. However, the short position L_s has little effect on the high 412 MHz resonant frequency in the simulation. Therefore, we can conclude that the C-shaped strip in Figure 1 mainly contributes to the low 391 MHz resonant frequency. Figure 4b also shows the effect of the feed position L_f on return loss. It is found that the longer the length L_f, the higher the low resonant frequency around 391 MHz. Similarly, the feed position hardly affects the resonant frequency of 412 MHz.

In order to further analyze the operation principle of the antenna, an antenna with different tuned strip length L_1 was examined. As shown in Figure 4c, the longer the length L_1, the lower the high resonant frequency about 412 MHz. However, the tuned strip length L_1 has almost no effect on the low

391 MHz resonant frequency. It is also worth pointing out that, according to simulation, the effect of L_2 on resonant frequency is similar to that of L_1. On the other hand, it can be concluded from Figure 3 that the total metal strip length contributes to the high resonant frequency, which is consistent with the conclusion of Figure 4c.

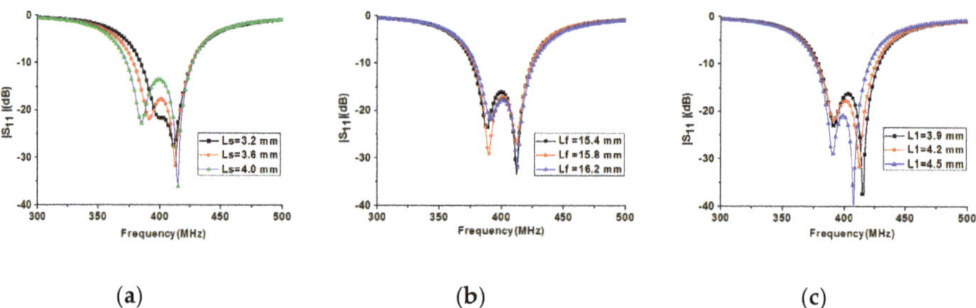

(a) **(b)** **(c)**

Figure 4. Comparison of $|S_{11}|$ of the proposed antenna with different geometric parameters: (**a**) short position L_s; (**b**) feed position L_f; and (**c**) tuned strip length L_1.

3. Results and Discussion

Photographs and the experimental setup of the fabricated antenna, including its superstrate, are shown in Figure 5. Figure 5a displays a photograph of the exploded flat antenna. The assembled flat and bent antennas with a 50 Ω coaxial cable are shown in Figure 5b,c, respectively. Figure 5d shows the experimental setup for the return loss measurement of the antenna in this study. According to [15], the permittivity and conductivity of chopped pork are similar to that of human skin in the MedRadio band, so the antenna was measured by using chopped pork. Figure 6 shows a comparison of measured and simulated return loss of the proposed antenna. The measured return loss agrees well with the simulation. The simulated -10 dB impedance bandwidth (IBW) is 49 MHz (378–427 MHz), and the measured $|S_{11}|$ is from 382 MHz to 434 MHz below -10 dB and has a bandwidth of 52 MHz, which completely covers the MedRadio band (401–406 MHz). Compared with the simulation, the slight frequency shift could be caused by unexpected fabrication tolerance and solder roughness.

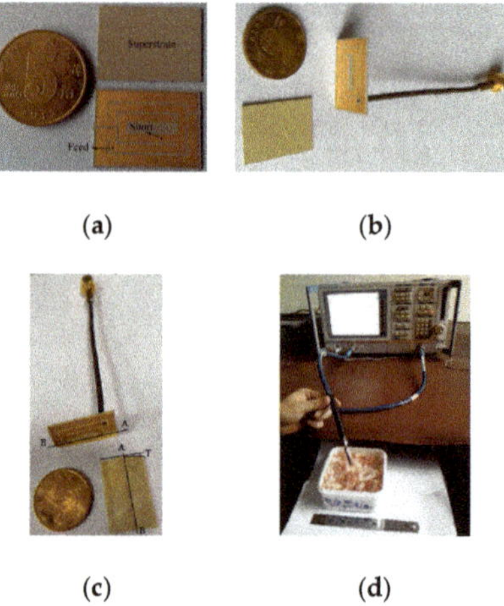

(a) **(b)**

(c) **(d)**

Figure 5. Photographs and experimental setup of the proposed antenna. (**a**) Photograph of the exploded flat antenna; (**b**) photograph of the assembled planar antenna; (**c**) photograph of the assembled bent antenna; (**d**) experimental setup.

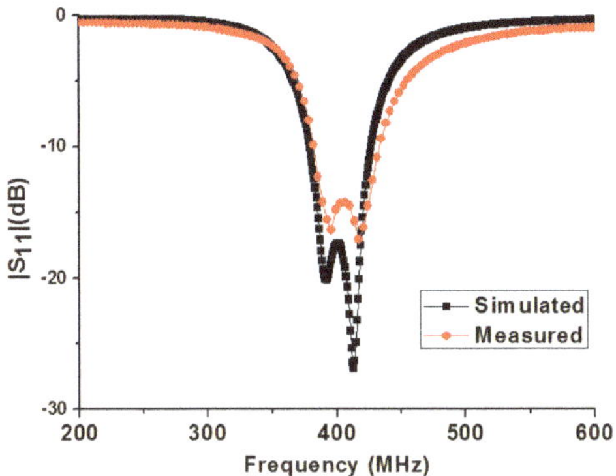

Figure 6. Measured and simulated $|S_{11}|$ of the proposed implantable antenna.

Typically, the implanted antenna is placed in a fixed position in the stationary human body. When the human body moves, bent and twisted antennas can be used to mimic it. In order to further analyze the performance of the proposed antenna in different cases, we compared the measured $|S_{11}|$ of the flat case and two bent cases with different bent angles $T = 10°$ and $T = 20°$ along the line AB in Figure 5c. Here, the experimental setup of Figure 5d is still employed. Measured results of the flat and bent antennas are shown in Figure 7. When the antenna is bent along the line AB, the high resonant frequency of the antenna is slightly lowered, while the low resonant frequency significantly increases, so the -10 dB impedance bandwidth is continuously reduced. However, it continues to completely cover the MedRadio band. It should be noted that the antenna cannot be arbitrarily distorted due to the limitation of the selected substrate in this work. By using ultra-thin and well-stretched substrates, related experiments will be conducted in the future.

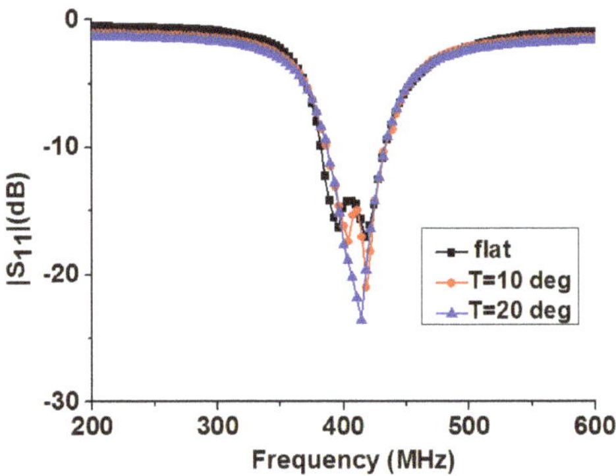

Figure 7. Measured $|S_{11}|$ of the flat and bent antenna with different angles.

The simulated two-dimensional far-field gain pattern of the proposed antenna at 402 MHz is shown in Figure 8. The pattern is nearly omni-directional on the YZ-plane with a peak gain of -34.9 dBi. In addition, we have evaluated the 1-g averaged specific absorption rate (SAR) for consideration of human body safety concerns [16]. When the delivered power of the proposed antenna is assumed to be 1 W, the maximum SAR value at 402 MHz is 284.5 W/kg. Therefore, the allowed transmitter power is 5.62 mW to satisfy the 1-g SAR regulation.

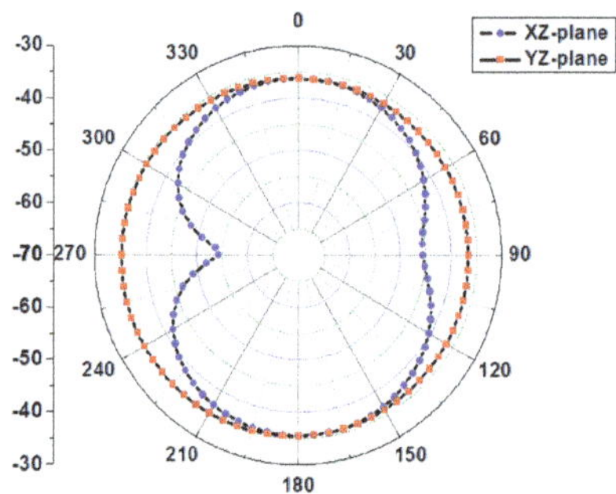

Figure 8. Simulated 2D patterns for the proposed implantable antenna at 402 MHz.

The size, -10 dB impedance bandwidth and peak gain of the proposed antenna are compared with those of previous broadband implantable antennas in the MICS or MedRadio bands in Table 1. From Table 1, we can see that our antenna has a wider bandwidth than other compact narrowband antennas [6,7]. In [8,9], complex stacked structures were used. A broadband monopole antenna with C-shaped coupled ground was adopted in [10], and has a slightly larger volume. In our work, by exciting dual resonant frequencies, a new implantable antenna with a single-layer radiating patch was proposed. Therefore, compared to similar broadband antennas [8–10], our antenna has good comprehensive performance in terms of structure, volume and bandwidth. It can be concluded that our antenna has a new structure and good comprehensive performance based on the results of the comparison.

Table 1. Performance comparison of the proposed antenna and other similar implantable antennas.

Reference	Volume (mm^3)	-10 dB Impedance Bandwidth (IBW) (MHz)	Peak Gain (dBi)
[6]	198	394–419	-32.4
[7]	151	398–423	-33.2
[8]	190	385–425	-26.0
[9]	791	353–473	-27.2
[10]	560	368–687	-28.0
This paper	479	378–427	-34.9

4. Conclusions

In this paper, we propose a broadband implantable antenna and discuss the bandwidth-increasing technology of implanted antennas. By using a bending and shorting strategy for a single-layer metal radiating patch, a novel dual-resonant implantable antenna in the MedRadio band has been presented. The proposed antenna has a 52-MHz measured bandwidth and a compact planar structure, without stacked metal radiating layers. The size of the proposed antenna is $23 \times 16.4 \times 1.27$ mm^3, which can be further reduced to meet the requirements of implantable microdevices by using a high dielectric constant and ultra-thin substrate. Good agreement is obtained between simulation and measurement for the return loss. In order to mimic human motion, the measured $|S_{11}|$ of the antenna in a flat and two bent cases is compared. Additionally, the radiation performance of the antenna and the human body safety performance were evaluated. The proposed antenna would be a promising candidate for implantable devices in the biotelemetry field owing to its advantages.

Author Contributions: R.L. provided the idea and built the model, B.L. performed computer simulations; R.L. and G.D. co-wrote the paper; X.S. reviewed the final manuscript; H.S. performed the experiments and analyzed the data.

Funding: This work was funded in part by the Industry-University Cooperative Education Project of Ministry of Education of China (grant number 201801238027), the Sichuan Provincial Science and Technology Key Project of China (grant numbers 19ZDYF0832, 2018GZ0261, 2017GZ0432), the Foundation of Chengdu University of Information Technology (grant number J201602), and the Innovation and Entrepreneurship Project of Chengdu University of Information Technology "Design of Array Antenna Compatible with GPS and Beidou Satellite Navigation System".

Conflicts of Interest: The authors declare no conflict of interest.

References

1. Greatbatch, W.; Homes, C.F. History of implantable devices. *IEEE Eng. Med. Biol. Mag.* **1991**, *10*, 38–41. [CrossRef] [PubMed]
2. Kiourti, A.; Nikita, K.S. A review of implantable patch antennas for biomedical telemetry challenges and solutions. *IEEE Trans. Antennas Propag.* **2012**, *54*, 210–228. [CrossRef]
3. Kiourti, A.; Nikita, K.S. Implantable antennas: A tutorial on design, fabrication, and in vitro/in vivo testing. *IEEE Microw. Mag.* **2014**, *15*, 77–91.
4. Kimi, J.; Rahmat-samii, Y. Plana inverted F antennas on implantable medical devices: Meandered type versus spiral type. *Microw. Opt. Technol. Lett.* **2006**, *48*, 567–572. [CrossRef]
5. Liu, C.R.; Guo, Y.X.; Xiao, S.Q. A hybrid patch/slot implantable antenna for biotelemetry devices. *IEEE Antennas Wirel. Propag. Lett.* **2012**, *11*, 1646–1649. [CrossRef]
6. Li, H.; Guo, Y.X.; Liu, C.R.; Xiao, S.Q. A miniature-implantable antenna for MedRadio-band biomedical telemetry. *IEEE Antennas Wirel. Propag. Lett.* **2015**, *14*, 1176–1179.
7. Li, R.Q.; Xiao, S.Q. Compact slotted semi-circular antenna for implantable medical devices. *Electron. Lett.* **2014**, *50*, 1675–1677. [CrossRef]
8. Liu, W.C.; Yeh, F.M.; Ghavami, M. Miniaturized implantable broadband antenna for biotelemetry communication. *Microw. Opt. Technol. Lett.* **2008**, *50*, 2407–2409. [CrossRef]
9. Lee, C.M.; Yo, T.C.; Huang, F.J.; Luo, C.H. Bandwidth enhancement of planar inverted-F antenna for implantable biotelemetry. *Microw. Opt. Technol. Lett.* **2009**, *51*, 749–752. [CrossRef]
10. Tsai, C.L.; Chen, K.W.; Yang, C.L. Implantable wideband low-SAR antenna with C-shape coupled ground. *IEEE Antennas Wirel. Propag. Lett.* **2015**, *14*, 1594–1597. [CrossRef]
11. Xu, L.J.; Guo, Y.X.; Wu, W. Bandwidth enhancement of an implantable antenna. *IEEE Antennas Wirel. Propag. Lett.* **2015**, *14*, 1510–1513. [CrossRef]
12. Alrawashdeh, R.S.; Huang, Y.; Kod, M.; Sajak, A.A.B. A Broadband Flexible Implantable Loop Antenna with Complementary Split Ring Resonators. *IEEE Antennas Wirel. Propag. Lett.* **2015**, *14*, 1506–1509. [CrossRef]
13. Das, S.; Mitra, D. A Compact Wideband Flexible Implantable Slot Antenna Design with Enhanced Gain. *IEEE Trans. Antennas Propag.* **2018**, *66*, 4309–4314. [CrossRef]
14. Karacolak, T.; Hood, A.Z.; Topsakal, E. Design of a dual-band implantable antenna and development of skin mimicking gels for continuous glucose monitoring. *IEEE Trans. Microw. Theory Tech.* **2008**, *56*, 1001–1008. [CrossRef]
15. Karacolak, T.; Cooper, R.; Unlu, E.S.; Topsaka, E. Dielectric of porcine skin tissue and in vivo testing of implantable antennas using pigs as model animals. *IEEE Antennas Wirel. Propag. Lett.* **2012**, *11*, 1686–1689. [CrossRef]
16. *IEEE Standard for Safety Levels with Respect to Human Exposure to Radio Frequency Electromagnetic Fields, 3 kHz to 300 GHz*; IEEE Standard C95. 1-1991; IEEE: Piscataway, NJ, USA, 1999.

Article

Numerical Study and Optimization of a Novel Piezoelectric Transducer for a Round-Window Stimulating Type Middle-Ear Implant

Houguang Liu [1], Hehe Wang [1,*], Zhushi Rao [2], Jianhua Yang [1] and Shanguo Yang [1]

[1] School of Mechatronic Engineering, China University of Mining and Technology, Xuzhou 221116, China; liuhg@cumt.edu.cn (H.L.); jianhuayang@cumt.edu.cn (J.Y.); ysgcumt@cumt.edu.cn (S.Y.)

[2] State Key Laboratory of Mechanical System and Vibrations, Shanghai Jiao Tong University, Shanghai 200240, China; zsrao@sjtu.edu.cn

* Correspondence: wanghehe@cumt.edu.cn; Tel.: +86-151-9067-0657

Received: 28 November 2018; Accepted: 4 January 2019; Published: 9 January 2019

Abstract: Round window (RW) stimulation is a new application of middle ear implants for treating hearing loss, especially for those with middle ear disease. However, most reports on it are based on the use of the floating mass transducer (FMT), which was not originally designed for round window stimulation. The mismatch of the FMT's diameter and the round window membrane's diameter and the uncontrollable preload of the transducer, leads to a high variability in its clinical outcomes. Accordingly, a new piezoelectric transducer for the round-window-stimulating-type middle ear implant is proposed in this paper. The transducer consists of a piezoelectric stack, a flextensional amplifier, a coupling rod, a salver, a plate, a titanium housing and a supporting spring. Based on a constructed coupling finite element model of the human ear and the transducer, the influences of the transducer design parameters on its performance were analyzed. The optimal structure of the supporting spring, which determines the transducer's resonance frequency, was ascertained. The results demonstrate that our designed transducer generates better output than the FMT, especially at low frequency. Besides this, the power consumption of the transducer was significantly decreased compared with a recently reported RW-stimulating piezoelectric transducer.

Keywords: piezoelectric transducer; middle ear implant; round window stimulation; hearing loss; flextensional amplifier; finite element analysis

---

1. Introduction

Hearing loss is one of the most common physical disabilities in our society, affecting more than 538 million people worldwide [1]. Despite impressive advances in microsurgical techniques, audiology and biotechnology, the relief we can offer to patients with hearing loss is still inadequate, especially for patients with sensorineural hearing loss [2]. Hearing aids can accomplish partial rehabilitation of sensorineural hearing loss; however, only a minority of patients choose to use them due to their inherent disadvantages, such as limited amplification, acoustic feedback, poor fidelity and the stigma of aging [3]. To overcome these shortcomings of conventional hearing aids, many institutions began to develop middle ear implants (MEIs) [4–8] which treat hearing loss directly by their implanted transducer's mechanical stimulation of the patient's ossicular chain. However, attaching the transducer to the ossicular chain is difficult in patients with middle ear disease, such as middle ear malformation, chronic otitis media and otosclerosis. To solve this issue, an alternative approach of coupling the transducer to the ear by stimulation of the round window (RW) membrane, called RW stimulation, has been widely investigated [9–11].

Although clinical data have confirmed the feasibility of RW stimulation, its outcomes have shown high variability. To determine the cause of these large variations, a number of studies have been conducted. Since the MED-EL Vibrant Soundbridge's floating mass transducer (FMT)—the transducer that has been widely used clinically for RW stimulation—has a diameter similar to the diameter of the RW, Nakajima et al. concluded that the bony overhang surrounding the RW could hinder the coupling of the FMT's motion to the RW [12]. Although introducing a compliant coupling layer between the transducer and the RW membrane would improve the vibration transmission to the RW [13], the coupling condition is difficult to control in surgery. Schraven et al. even stated that the high variability in the postoperative outcome of RW stimulation is primarily owing to the success of this coupling [14]. Besides this, Maier et al. performed an experimental study on human temporal bone which indicated a clear dependency of the sound transmission efficiency on the static force applied to the RW membrane [15]. Furthermore, based on an experimental investigation, Muller et al. suggested that the optimal preload of the transducer should less than 20 mN [16]. However, as most transducers used in the clinic are not originally designed for RW stimulation, the preload of the transducer on the round window membrane (RWM) cannot be precisely monitored.

In addition to high output variability, reports also show that using the FMT for RW stimulation has a considerably low output at lower frequencies (< 1 kHz) [17,18]. To address this issue, Shin et al. designed a novel piezoelectric transducer specifically for RW stimulation [19]. In comparison with the widely used electromagnetic transducer, that is, the FMT, their experimental results demonstrated that their transducer has excellent low-frequency output. Moreover, this piezoelectric transducer displays the advantages of ease of fabrication, wider bandwidth and compatibility with an external magnetic environment. However, the preload of the transducer applied to the RW membrane is still difficult to precisely control.

Accordingly, in this paper, we propose a new piezoelectric transducer incorporating a preload visual indicator for RW stimulation. Besides this, to further decrease the power consumption of the piezoelectric transducer, we introduced a flextensional amplifier for the piezoelectric stack in the transducer. Then, based on a coupling finite element model of a human ear and the transducer, the main design parameters of the transducer were optimized to produce sufficient driving force for RW stimulation. The results indicate that the new piezoelectric transducer not only has excellent output over the audio frequency range but also significantly decreases the power consumption.

2. Materials and Methods

2.1. Concept and Structure of the Piezoelectric Transducer

Figure 1 shows a schematic view of the proposed round-window-stimulating-type middle ear implant (RW-MEI) in the human ear. This RW-MEI mainly consists of three parts: a piezoelectric transducer, a microphone and an assembly of a signal processing module and a rechargeable battery. The microphone, which is implanted near the entrance of the ear canal to take advantage of the natural sound filtering from the pinna, is responsible for picking up sound from the outside and generating a corresponding electrical signal. Then, the signal is transmitted to the signal processing module, which processes it according to the patient's individual hearing loss conditions and produces a driving electrical signal for the implanted piezoelectric transducer. Finally, the piezoelectric transducer, which is placed in the middle ear cavity with its tip attached to the round window membrane, converts the electrical signal into a mechanical vibration that is delivered directly to the cochlea. In the cochlea, the vibration is detected by hair cells and perceived as sound.

The inner structure of the round-window-stimulating-type piezoelectric transducer is shown in Figure 2. This transducer is composed of a coupling rod, a salver, a piezoelectric component, a titanium housing, a plate and a supporting spring. The piezoelectric component consists of two parts: a piezoelectric stack and a flextensional amplifier. The flextensional amplifier was introduced to amplify the displacement output of the piezoelectric stack, so as to decrease the required driving voltage and reduce the power consumption. To enable the surgeon to detect the preload applied on the RW membrane, a force visual indicator was set on the supporting spring. During the surgical procedure, the transducer is implanted into a bone bed drilled behind the mastoid. The back end of the supporting spring is contacted with the bony wall opposite the RW membrane. The other end of the supporting spring fixes to a plate, which connects with the base side of the piezoelectric components. Biocompatibility of the piezoelectric component is secured by coating its entire surface with biocompatible materials and putting it into a titanium housing. Besides this, according to an experimental study by Arnold et al. [20], the vibration transmission to the cochlea is best when the transducer output is perpendicular to the round window membrane. To ensure the direction of the transducer output, we fixed the transducer's output coupling rod on a salver, which is settled in the titanium housing and connected the salver with the piezoelectric component. The geometry of the salver's outer wall is consistent with that of the titanium housing's inner wall, thereby forming a sliding pair and guiding the output direction of the piezoelectric component along the longitudinal direction of the titanium housing. The tip of the coupling rod attaches to the round window membrane and transmits the transducer's vibration to the cochlea.

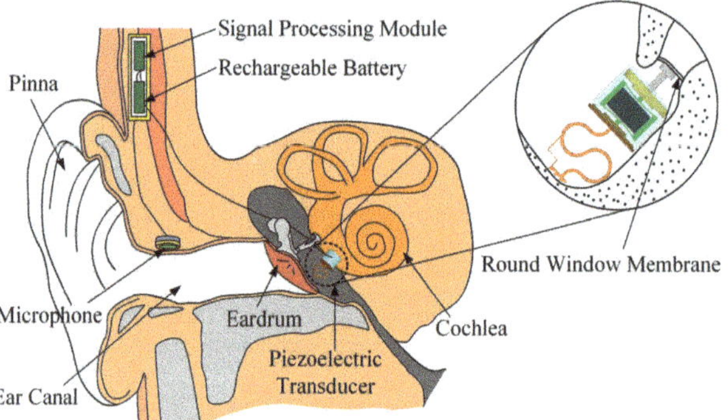

Figure 1. Concept illustration of the round window (RW)-stimulating-type middle ear implant.

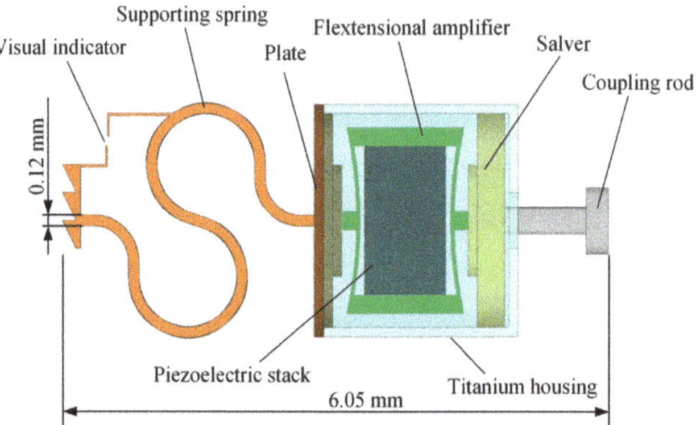

Figure 2. The inner structure of the proposed piezoelectric transducer.

2.2. Design of the Piezoelectric Transducer

2.2.1. Design of the Piezoelectric Component

The piezoelectric component of this transducer consists of a piezoelectric stack and a flextensional amplifier, as shown in Figure 3. Considering the small space in the human middle ear cavity, where the transducer is implanted, the size of the transducer should be restricted. For comparison, the piezoelectric stack used in this study is the same as that used by Shin et al. [19]. Specifically, this piezoelectric stack is made of lead zirconate titanate ceramics (PZT-8) with dimensions of 0.9 mm × 0.9 mm × 1.6 mm, which is small enough for this application. This piezoelectric stack has a layer number of 222 to produce 300 nm displacement under a voltage of 6 V. The flextensional amplifier was incorporated to further decrease the transducer's power consumption. As shown in Figure 3, once the piezoelectric stack generates a Δx deformation driven by a voltage, the flextensional amplifier would generate an amplified output displacement: Δy. Their relationship can be expressed as Equation (1), where G is the displacement amplification ratio of the flextensional amplifier. The value of G can be estimated based on Equation (2):

$$\Delta y = G \Delta x \tag{1}$$

$$G = L/2D \tag{2}$$

where L is the length of the amplifier's inner cavity and D is the amplifier's unilateral inner concave depth. Considering the spatial limitation in the middle ear cavity, the height (H), the width (W) and the thickness of the amplifier were set to 1.3 mm, 2.0 mm and 0.9 mm, respectively. L and D were set to 1.6 mm and 0.15 mm. The thickness of the flexible arm was set to 42 µm. According to Equation (2), the obtained amplification ratio is 5.3.

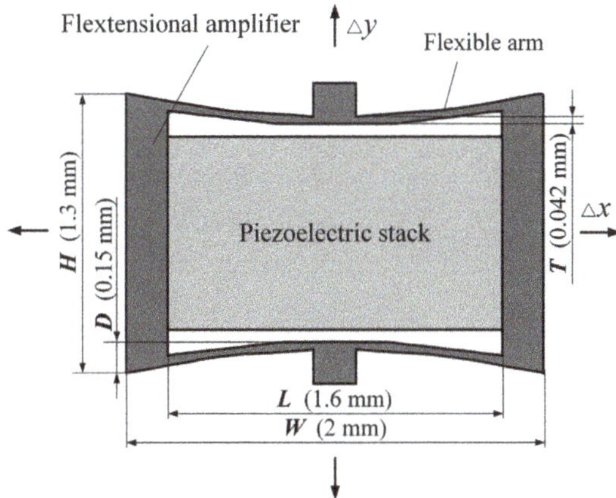

Figure 3. The structure of the piezoelectric component.

2.2.2. Design of the Supporting Spring

The design of the supporting spring significantly influences the transducer's resonant frequency. Considering that the middle ear implant is designed for compensating hearing loss, its frequency characteristics should be similar to those of the human middle ear. Based on a human temporal bone experiment, Homma et al. reported that the human middle ear resonant frequency is around 0.8–1.2 kHz [21]; therefore, the resonant frequency of our designed transducer should lie within this frequency range.

To facilitate the design of this supporting spring, we constructed a finite element model of the piezoelectric transducer, as shown in Figure 4. This finite element model consists of six components: the coupling rod, the salver, the flextensional amplifier, the plate, the titanium housing and the supporting spring. The piezoelectric stack, which is made of PZT-8, was meshed by 1600 hexahedral elements. The density of PZT-8 is 7600 kg/m³. The other material properties of it are listed in Table 1. The piezoelectric finite element model equations can be written in terms of the nodal displacement U

and nodal electrical potential Φ for each node. Under the action of the nodal electric charge Q and the external mechanical forces F, the equilibrium equations are expressed in matrix form as [22]

$$\begin{bmatrix} M_{uu} & 0 \\ 0 & 0 \end{bmatrix} \begin{bmatrix} \ddot{U} \\ \ddot{\Phi} \end{bmatrix} + \begin{bmatrix} C_{uu} & 0 \\ 0 & 0 \end{bmatrix} \begin{bmatrix} \dot{U} \\ \dot{\Phi} \end{bmatrix} + \begin{bmatrix} K_{uu} & K_{u\varphi} \\ K_{u\varphi}{}^{\mathrm{T}} & K_{\varphi\varphi} \end{bmatrix} \begin{bmatrix} U \\ \Phi \end{bmatrix} = \begin{bmatrix} F \\ Q \end{bmatrix} \tag{3}$$

with

$$K_{uu} = \iiint_{\Omega e} B_u{}^{\mathrm{T}} c B_u \mathrm{d}V, \; K_{u\varphi} = \iiint_{\Omega e} B_u{}^{\mathrm{T}} e B_\varphi \mathrm{d}V, \; K_{\varphi\varphi} = \iiint_{\Omega e} B_\varphi{}^{\mathrm{T}} \varepsilon B_\varphi \mathrm{d}V$$

$$M_{uu} = \rho \iiint_{\Omega e} N_u{}^{\mathrm{T}} N_u \mathrm{d}V, \; C_{uu} = f i K_{uu}$$

denoting the mechanical stiffness matrix, piezoelectric coupling matrix, dielectric stiffness matrix, mass matrix and mechanical damping matrix, respectively.

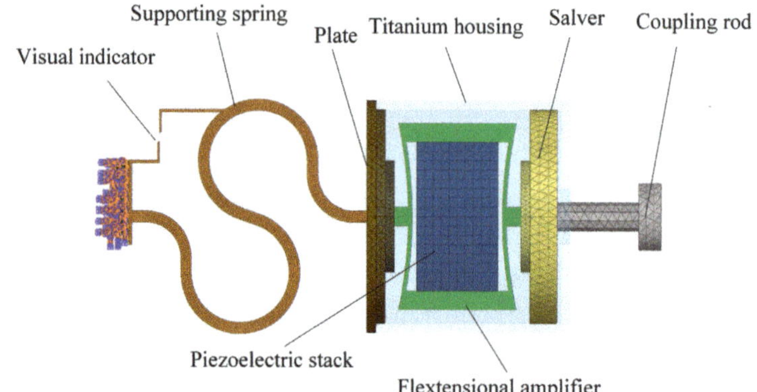

Figure 4. Finite element model of the proposed piezoelectric transducer.

Table 1. Material parameters of lead zirconate titanate ceramics (PZT-8).

Elastic Stiffness Constant (GN/m^2)						Piezoelectric Constant (C/m^2)			Permittivity Constant ($\times 10^{-10}$ F/m)	
c_{11}^E	c_{12}^E	c_{13}^E	c_{33}^E	c_{44}^E	c_{66}^E	e_{15}	e_{31}	e_{33}	ε_{11}^s	ε_{33}^s
146.9	81.1	81.1	131.7	31.4	32.9	10.3	−3.9	14.0	114.2	88.5

All the other parts of the transducer are made of titanium. The corresponding density is 4430 kg/m^3, the Young's modulus is 116 GPa and the Poisson's ratio is 0.32. The flextensional amplifier was meshed by 7680 hexahedral elements. The salver, titanium housing, plate and supporting spring were meshed by 15224, 13076, 13228 and 21144 tetrahedral elements, respectively. The connections between the coupling rod and salver, the salver and the flextensional amplifier, the flextensional amplifier and plate and the plate and the supporting spring were modeled by coupling the corresponding finite element nodes of them. The end of the supporting spring was fixed by defining a fixed constraint on the corresponding nodes. The influence of the supporting spring's cross-sectional area on the transducer's frequency characteristics was analyzed by changing the width of its cross section (L in Figure 5). The model-predicted transducer output velocity is shown in Figure 6, in which the transducer was driven by a voltage of 0.5 V. It shows that when the cross-sectional area of the supporting spring changes from 0.024 mm^2 to 0.036 mm^2, 0.048 mm^2, or 0.060 mm^2, the transducer's resonant frequency changes from 788 Hz to 938 Hz, 1082 Hz, or 1243 Hz, respectively. To make sure that the transducer has a resonant frequency close to 1100 Hz, we selected 0.048 mm^2 as the cross-sectional area of the supporting spring. Thus, the final width of the supporting spring was

0.4 mm. Based on this geometric size, the static analysis showed that the visual indicator deformed 20 µm under 20 mN; therefore, the width of the two bars on the visual indicator was set to 20 µm to make them coincide when the transducer preload increases to 20 mN.

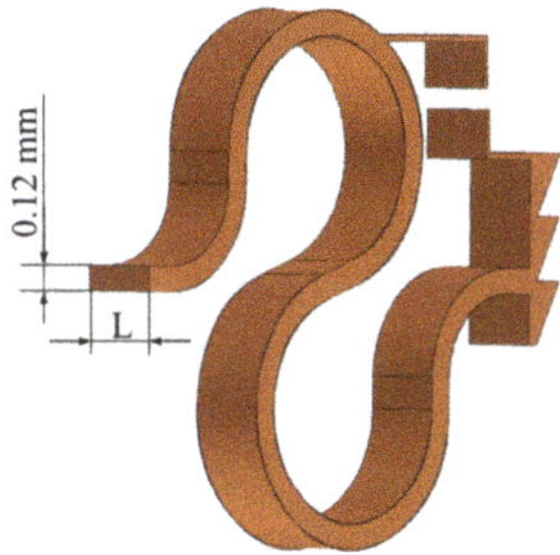

Figure 5. The structure of the supporting spring.

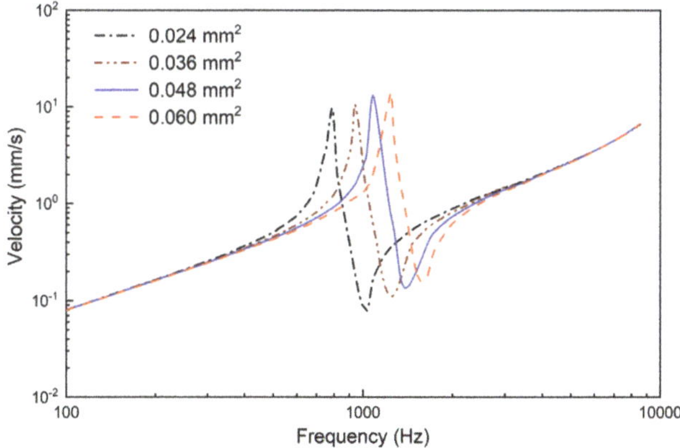

Figure 6. The effect of the supporting spring's cross-sectional area on the output of the transducer.

2.2.3. Design of the Coupling Rod Tip

Experimental and theoretical studies have shown that the geometry of the transducer's tip, which connects the round window membrane and transmits vibration to the cochlea, influences the transducer's hearing compensation performance [13,23]. To facilitate the design of our coupling rod's tip, we utilized our previously established human ear finite element model (as shown in Figure 7), which is composed of the middle ear and cochlea. In this paper, we modified the Young's modulus of some tissues according to literature reports to improve the reliability of this model. The changed parameters and their referenced sources are listed in Table 2. To confirm the validity of the model, we carried out three sets of comparisons with reported experimental results. Firstly, considering that the stapes is responsible for transmitting sound into the cochlea, we calculated its footplate displacement in our model and compared it with Gan et al.'s data [24], as shown in Figure 8. Then, the cochlear input impedance, which also reflects the sound transfer property from the middle ear into the cochlea, was also selected to verify our model. The model-predicted result was plotted with published experimental reports [25–27], as shown in Figure 9. Finally, the excited response at a specific place (12 mm from the stapes) on the basilar membrane was calculated and compared with Stenfelt et al. [28] and Gundersen et al.'s [29] measurement data (Figure 10). These comparisons show that our model-predicted results are generally consistent with the experimental data. Thus, we can use this model to aid in the design of this transducer.

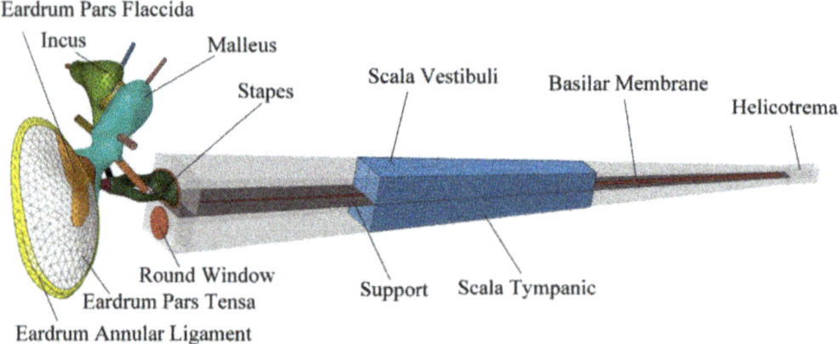

Figure 7. The constructed human ear finite element model.

Table 2. The modified Young's modulus values in our human ear finite element model.

Components	Young's Modulus, N/m^2	Ref.
Incudostapedial joint	4.4×10^5	[30]
Stapedial annular ligament	1.5×10^4	[31]
Lateral mallear ligament	6.7×10^4	[31]
Superior mallear ligament	4.9×10^4	[31]

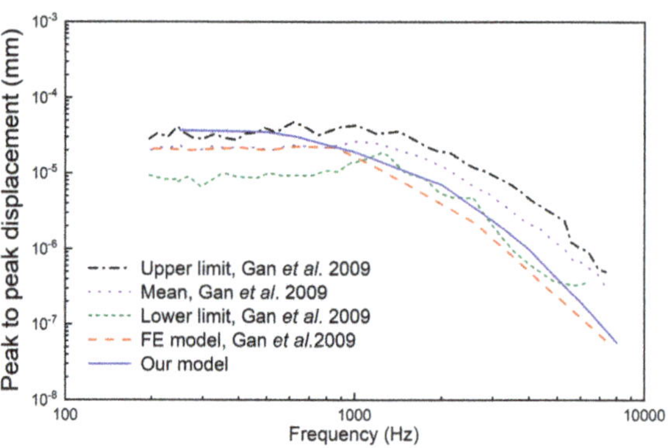

Figure 8. Comparison of stapes footplate displacement under 90 dB sound pressure level (SPL) applied at the eardrum.

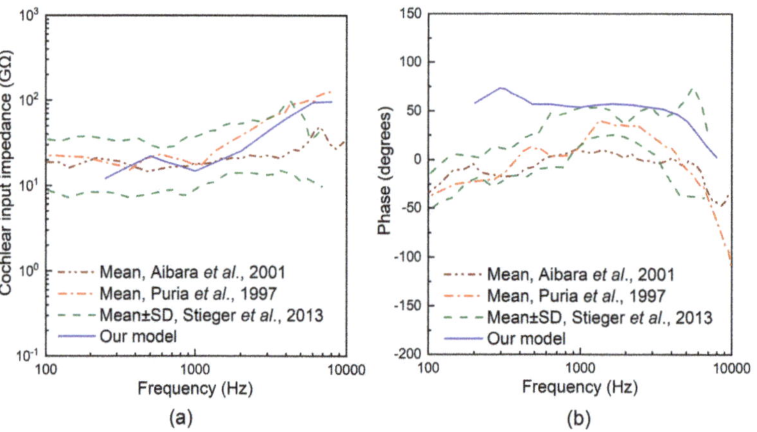

Figure 9. Comparison of the cochlear input impedance: (**a**) magnitude; (**b**) phase.

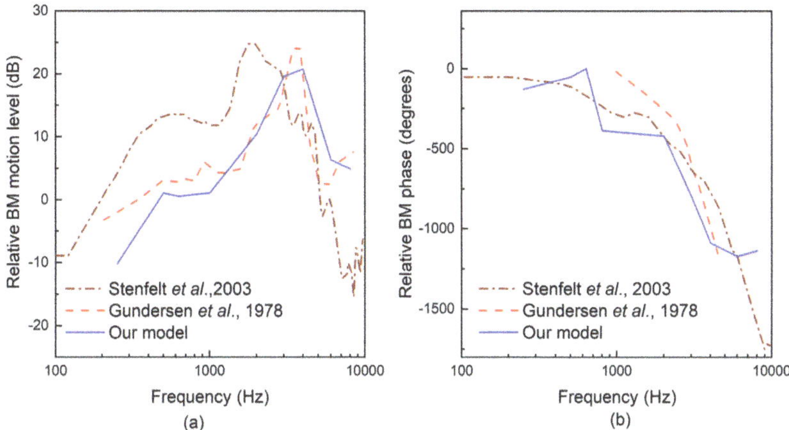

Figure 10. Comparison of the relative basilar membrane motion level at 12 mm from the stapes: (**a**) magnitude; (**b**) phase.

Based on this verified human ear finite element model, we constructed a coupling mechanical model of the human ear and the piezoelectric transducer by coupling the nodes of the transducer's tip with the attached nodes on the round window membrane. The final constructed coupling model is shown in Figure 11. Then, we changed the coupling rod tip's geometry and size to study the effect of these parameters on the transducer's performance. As shown in Figure 12, the following five tips were analyzed: (a) a 0.5 mm diameter spherical tip; (b) a 0.5 mm diameter cylindrical tip; (c) a cubic tip with a 0.44 mm × 0.44 mm square top surface; (d) a 0.72 mm diameter cylindrical tip; and (e) a cubic tip with a 0.64 mm × 0.64 mm square top surface. The influence of these tips on the transducer's performance is shown in Figure 13. It demonstrates that, with the same diameter, the transducer with the cylindrical tip can stimulate the stapes footplate velocity more than can the transducer with the spherical tip. However, with the same cross-sectional area, the performance of the transducer with the cylindrical tip is similar to that with the cubic tip. Besides this, when the transducer's tip cross-sectional area increased from 0.2 mm^2 to 0.4 mm^2, the transducer-stimulated stapes footplate velocity increased significantly, especially at high frequency. Considering that most sensorineural hearing loss is severe at high frequency and that the cubic tip is more likely to contact the bony surrounding of the round window, we selected the 0.72 mm diameter cylindrical tip, which has a cross-sectional area accounting for 20% of the RW membrane's cross-sectional area, as our transducer coupling rod's tip.

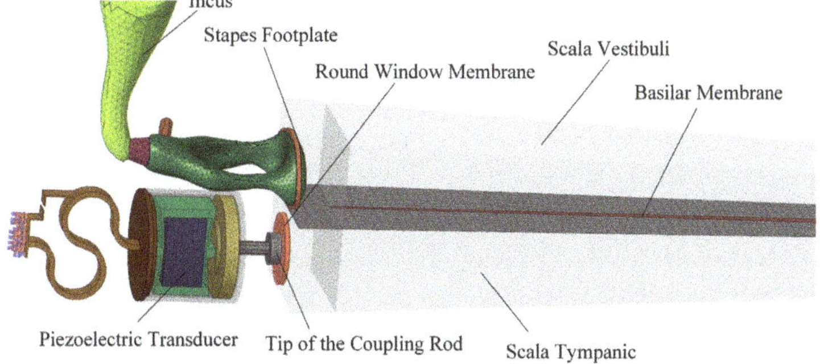

Figure 11. Constructed coupling model of the human ear and the piezoelectric transducer.

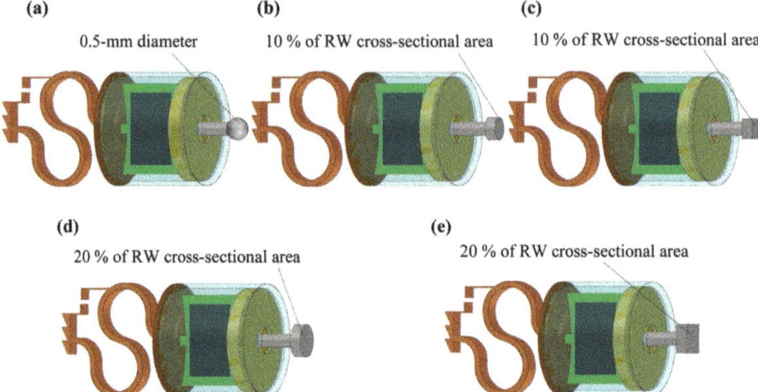

Figure 12. The studied transducer coupling rod's tips with different geometries: (**a**) 0.5 mm diameter spherical tip; (**b**) 0.5 mm diameter cylindrical tip; (**c**) cubic tip with a 0.44 mm × 0.44 mm square top surface; (**d**) 0.72 mm diameter cylindrical tip; (**e**) cubic tip with a 0.64 mm × 0.64 mm square top surface.

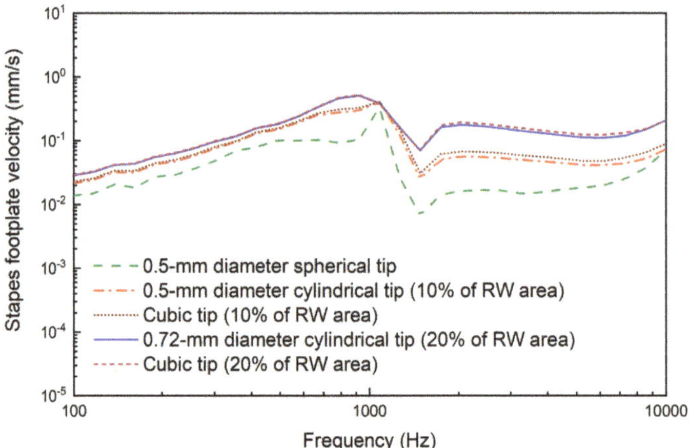

Figure 13. The influence of the coupling rod tip's geometry on the transducer's performance (at 0.5 V).

2.3. Evaluation of the Transducer's Performance

To evaluate the hearing compensating performance of our designed transducer, we calculated the stapes velocity and compared it with those driven by 94 dB sound pressure level (SPL) sound stimulation applied to the eardrum. Then, the transducer-excited response was converted to the equivalent sound pressure level at the eardrum to generate the same response of the stapes. The corresponding equivalent sound pressure (ESP) of the transducer's stimulating performance was calculated according to Equation (4):

$$P_{eq} = 94 + 20 \log_{10}\left(\frac{v_{tr}}{v_{ac}}\right) \tag{4}$$

where v_{ac} is the sound-stimulated (94 dB SPL) velocity of the stapes and v_{tr} is the corresponding transducer-stimulated velocity of the stapes.

3. Results

The model-predicted stapes footplate velocity stimulated by the piezoelectric transducer is shown in Figure 14. For comparison, the stapes footplate velocity excited by 94 dB sound pressure applied at the eardrum was also plotted. It demonstrates that the stapes velocity stimulated by this piezoelectric transducer at 0.5 V is similar to that under eardrum acoustical stimulation at 94 dB SPL. Besides this, at high frequency, the piezoelectric-transducer-stimulated stapes vibration is considerably larger than that by acoustical stimulation.

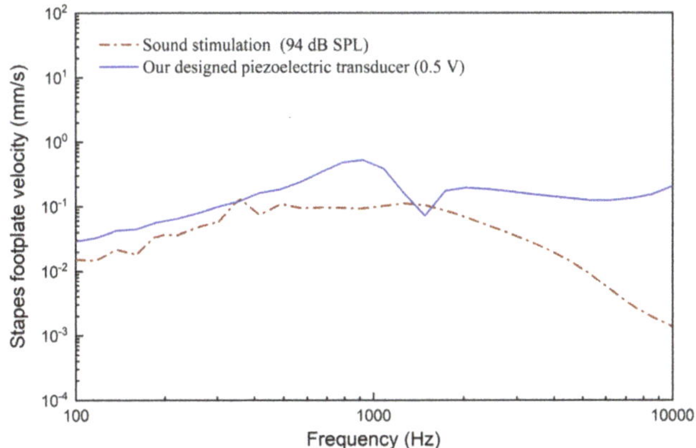

Figure 14. Stapes vibration stimulated by the piezoelectric transducer (0.5 V) and acoustic stimulations (94 dB SPL at the eardrum).

Figure 15 shows the equivalent sound pressure excited by the piezoelectric transducer at 0.5 V. It indicates that the piezoelectric transducer driven by 0.5 V can generate about 100 dB SPL equivalent sound pressure at the eardrum below 3 kHz, 110 dB at 4 kHz and 120 dB above 6 kHz. Considering many middle ear implant studies use 100 dB SPL as the design criterion, this piezoelectric transducer's output is adequate. Meanwhile, it also demonstrates that the transducer can compensate severe hearing loss at high frequency, which is a valuable advantage given that most sensorineural hearing loss deteriorates high frequencies more severely than it does low frequencies.

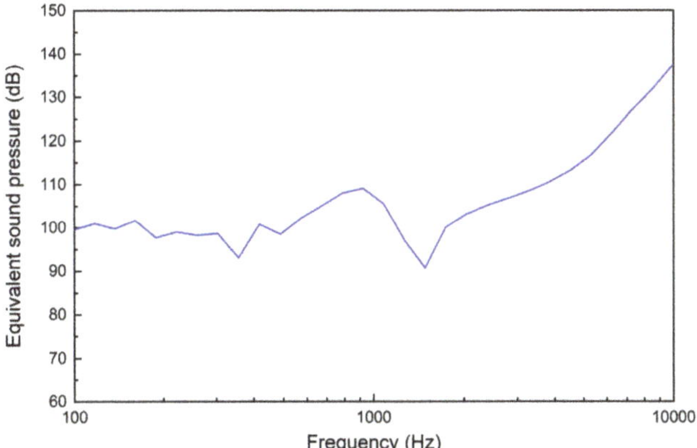

Figure 15. Equivalent sound pressure for the piezoelectric transducer stimulation at 0.5 V.

4. Discussion

Since human ear is a biology system with a complex structure and ultra-small dimension, systematic experimental study on it for the transducer is difficult to carry out. To aid the design of the transducers in the middle ear implants, many studies utilized human ear finite element models [32–34]. Their experimental results have confirmed the validation of this method. Thus, in this paper, we take the same method to optimize our transducer. Furthermore, to ensure the accuracy of our model, we constructed our human ear finite element (FE) model based on high-resolution micro-CT scanning images of fresh human temporal bone. Besides this, we further carried out three sets of comparisons with reported experimental data to verify our FE model's reliability. Comparison results show that our human ear FE model is reliable and can be used to aid the design of the transducer of middle ear implant.

To treat sensorineural hearing loss in the presence of middle ear disease, in which a transducer cannot couple to the ossicular chain, MED-EL Vibrant Soundbridge's FMT has been widely implanted on the round window membrane clinically. However, the FMT was not originally designed for RW stimulation. There exists a geometric mismatch between the FMT (with a diameter of 1.8 mm) and the round window membrane (with a diameter of 1.5–1.9 mm [35]). This geometric mismatch would cause the transducer to attach to the bony surrounding of the round window and decrease its transmission efficiency to the cochlea. The tip of our transducer specifically designed for RW stimulation has a diameter of 0.72 mm. This diameter is much less than that of the round window membrane and guarantees the stability of the transducer's performance. Our study also shows that the piezoelectric transducer's transmission efficiency is proportional to the cross-sectional area of the transducer's tip. This result is consistent with that of the experimental report conducted by Schraven et al. [23].

The preload of the transducer on the round window membrane also affects the transducer's performance and leads to variability in patients' postoperative outcomes [15,18]. Based on a human temporal bone experiment, Mathias further suggested that the optimal preload of the transducer should be less than 20 mN [16]. However, it is difficult to control the preload during surgery for FMT implantation, since the FMT is fixed just by putting cartilage or connective tissue behind it. Our piezoelectric transducer introduces a force visual indicator with two bars which would coincide when the applied force equals 20 mN. This force visual indicator can help the surgeon to monitor the preload of the transducer during the operation. Besides this, Ishii et al. found that a force of 564 mN applied on the round window membrane would rupture the membrane [36] and lead to further hearing loss. Therefore, the incorporated force visual indicator can also assist the surgeon to prevent the round window membrane's rupture and protect the patient's residual hearing.

Hearing loss compensating capability, which should meet the patient's requirement, is a crucial criterion for middle ear implant design. As previously noted, RW stimulation is a new application of middle ear implants for treating sensorineural hearing loss in the presence of middle ear disease. Since most sensorineural hearing loss is severe at high frequency [2], the transducer should have good performance at high frequency. Besides this, as middle ear diseases (e.g., otosclerosis) are usually worse at low frequency, low-frequency capability is also important for round window stimulation. However, the widely used RW-stimulating transducer, that is, FMT, is designed for stimulating the incus long process and treating sensorineural hearing loss. Clinical reports show that it has a poor output at low frequency [10,37]. In order to study the performance of our designed transducer comparatively, a comparison between our transducer-stimulated stapes velocity with that driven by the FMT is made in Figure 16. In this figure, the FMT-stimulated results were obtained from Nakajima et al.'s experimental data [12], which were measured in the human temporal bone. For comparison, the acoustical-stimulation-excited (94dB SPL) stapes velocities taken from the ASTM-F2504 standard [38] were also plotted in Figure 16. It shows that the FMT-simulated stapes response is similar to that excited by acoustical stimulation at high frequency (>1000 Hz) but significantly smaller at frequencies lower than 1000 Hz, especially at 250 Hz. The poor low-frequency performance would lower patients' speech perception (vowel sounds) and music perception. In contrast, our designed transducer not only provides a superior output at high frequency but also generates sufficient stapes response at low frequency. This characteristic would benefit the compensation of mixed hearing loss, which has been encountered in the clinical practice of RW stimulation.

The performance of the novel piezoelectric transducer specifically designed for round window stimulation by Shin et al. [19] was also selected for comparison. As shown in Figure 16, compared with the FMT, Shin et al.'s piezoelectric transducer also has an excellent low-frequency output. The stapes velocities driven by Shin et al.'s piezoelectric transducer is similar to those excited by normal acoustical stimulation at low frequency (<1000 Hz). Moreover, Shin et al.'s transducer has a better performance at high frequency. Compared with Shin et al.'s results, our designed piezoelectric transducer can stimulate greater stapes response at low frequency. However, it should be noted that the stapes velocities plotted in Figure 16 were derived from different human temporal bones. Thus, comparison of only the transducer-stimulated stapes velocities neglects the individual influence of the temporal bone used. To address this issue, we further calculated the equivalent sound pressure based on the corresponding temporal bone's acoustical stimulation result, as shown in Figure 17. It demonstrates that under a 0.5 V driving voltage, our piezoelectric transducer can generate an ESP similar to that driven by Shin et al.'s transducer at 6 V.

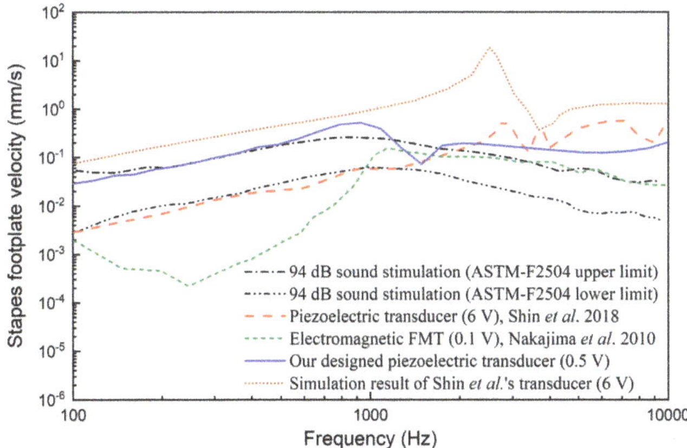

Figure 16. Comparison of stapes responses driven by our piezoelectric transducer and reported transducers.

The above reduction of the required driving voltage is mainly attributed to the improved fixation method of our transducer. Shin et al.'s transducer was fixed by placing dental resin cement on the side of their transducer's housing; therefore, the coupling condition of the transducer with the round window membrane is difficult to control. Similar to Shin et al.'s transducer, the coupling condition of the FMT is also difficult to control. Investigations of different coupling conditions with FMT in temporal bones, Arnold et al. found that variability between individual results was about 40 dB when using standard-fixation method [39]. To figure out the influence of the transducer's coupling condition in Shin et al's experiment, we calculated the ideal stapes footplate velocity excited by Shin et al.'s transducer using our human ear FE model. Since their transducer's diameter (1.75 mm) is bigger than the diameter of our model's round window membrane (1.6 mm), we cannot construct a coupling model of their transducer and our human ear FE model. Considering that an ideal piezoelectric transducer is a displacement-driven transducer [40], we stimulated our model's round window membrane directly using their transducer's output displacement. The model-predicted results are also plotted in Figures 16 and 17. They show that the simulation result for Shin et al.'s transducer is significantly larger than their experimental result. The fixation method of their transducer introduced a deterioration of 5 dB (at 375 Hz) to 24 dB (at 1395 Hz) in the equivalent sound pressure, which is consistent with Arnold et al.'s results [39].

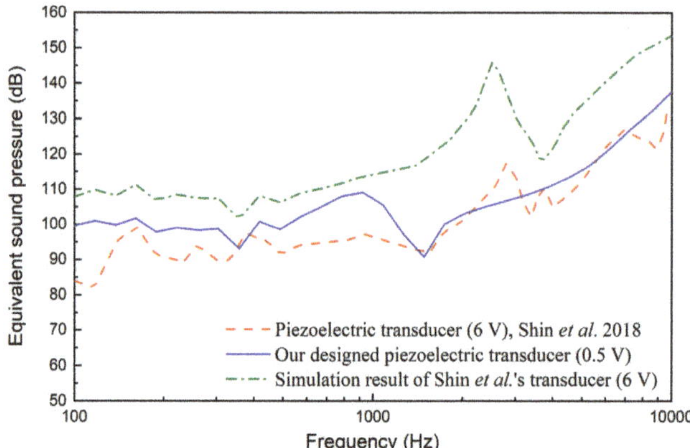

Figure 17. Comparison of the equivalent sound pressures of our piezoelectric transducer and reported transducer.

In addition to the improved fixation method of the transducer, the introduction of the flextensional amplifier in our transducer also reduced the required driving voltage and boosted the hearing loss compensating performance. To analyze this improvement, we compared the simulated stapes footplate velocities driven by our transducer and Shin et al.'s transducer under the same driving voltage (6 V). Under this simulation, the coupling condition's influence of Shin et al.'s transducer was excluded by applied their transducer's output directly on the round window membrane of our model. The calculated gain in our transducer-stimulated velocity relative to that by Shin et al.'s transducer is shown in Figure 18. It demonstrates that our transducer did amplify the stimulated stapes velocity, especially at low frequency (<1 kHz), with an amplification gain of about 5. This amplification gain value is close to that (5.3) which we aimed for during the design of the flextensional amplifier (in Section 2.2.1).

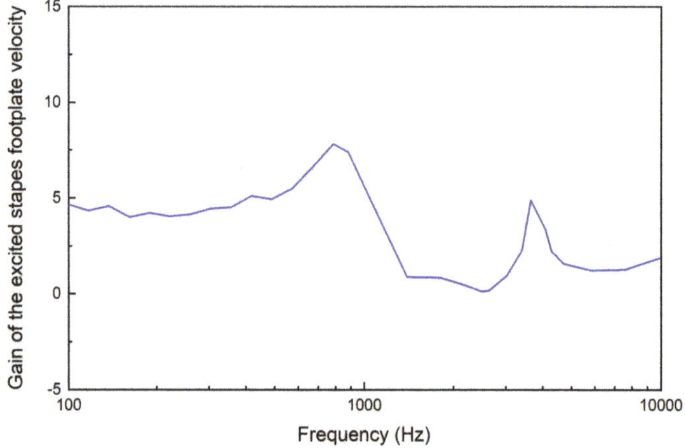

Figure 18. The gain in our transducer-stimulated velocity relative to that stimulated by Shin et al.'s transducer (6 V) [19].

A middle ear implant is a device implanted in the human ear; therefore, its transducer's power consumption should be restricted. The piezoelectric transducer's behavior can be approximated to that of a capacitor when its working frequency is far less than its resonant frequency. As shown in Equation (5), the piezoelectric transducer's power consumption can be calculated as [5]

$$P_{\text{rms}} = I_{\text{rms}} V_{\text{rms}} \cos\theta = 2\pi f C V_{\text{rms}}^2 \cos\theta \tag{5}$$

where V_{rms} is the driving voltage, f is the frequency, C is the capacitance of the transducer's piezoelectric stack and $\cos\theta$ is the power factor. Considering that our piezoelectric stack is the same one used by Shin et al., the decrease in the transducer's power consumption is proportional to the square of the voltage increase. Therefore, the reduction of our designed transducer's working voltage will significantly decrease its power consumption, which is an advantage for a totally implanted hearing device.

Compared with FMT, our transducer's structure is relatively complicated. This because of the fact that the FMT was not designed specifically for the RW stimulation; therefore, it does not contain the fixation part fixing the device against the round window and the coupling rod transmitting its vibration to the round window membrane. Nevertheless, compared with other commercial piezoelectric transducers for the middle ear implants, for example, Envoy Medical Corporation's Esteem [41], our transducer's mechanical structure is still simple. The key component of our transducer is the piezoelectric component as it responsible for converting the electrical signal to vibration and stimulating the round window membrane. The piezoelectric component consists of two parts: the flextensional amplifier and the piezoelectric stack. The piezoelectric stack, which is used in our transducer, is a commercial product (PAZ-10-0079) produced by Murata Manufacturing Co., Ltd. (Kyoto, Japan). In terms of the flextensional amplifier, its dimensions are compatible to that of the flextensional amplifier fabricated by Wang et al. [42]. Therefore, the mechanical structure of the piezoelectric transducer is acceptable in terms of fabrication.

5. Conclusions

Currently, the transducer widely used for round window stimulation (i.e., FMT) has three inherent shortcomings: mismatch between the diameter of the transducer and the RW membrane, uncontrollable preload and poor low-frequency output. To overcome these problems, we proposed a new piezoelectric transducer specifically for round window stimulation. To facilitate the design process of this transducer, a coupling finite element model of the transducer and a human ear was constructed. Then, the transducer's key design parameters were optimized based on this model. The transducer's hearing loss compensating performance was also studied and compared with those of the FMT and a recently reported piezoelectric transducer. The results show that our designed transducer generates better output than the FMT, especially at low frequency. Besides this, the power consumption of the transducer was significantly decreased compared with that of a recently reported RW-stimulating piezoelectric transducer.

In subsequent research, we will fabricate the piezoelectric transducer and carry out a human temporal bone experiment to evaluate its feasibility.

Author Contributions: Conceptualization, H.L.; Methodology, H.L., H.W., Z.R. and J.Y.; Software, H.L., H.W. and Z.R.; Validation, H.L., S.Y.; Formal Analysis, H.L., H.W.; Original Draft Preparation, H.L.

Funding: This research was funded by National Natural Science Foundation of China (grant number 51775547) and the Top-notch Academic Programs Project of Jiangsu Higher Education Institutions (grant number PPZY2015B120).

Conflicts of Interest: The authors declare no conflict of interest.

References

1. Stevens, G.; Flaxman, S.; Brunskill, E.; Mascarenhas, M.; Mathers, C.D.; Finucane, M. Global Burden of Disease Hearing Loss Expert Group. Global and regional hearing impairment prevalence: An analysis of 42 studies in 29 countries. *Eur. J. Public Health* **2013**, *23*, 146–152. [CrossRef] [PubMed]
2. Moore, B.C.J.; ebrary Inc. *Cochlear Hearing Loss Physiological, Psychological and Technical Issues*, 2nd ed.; John Wiley & Sons: Chichester, UK; Hoboken, NJ, USA, 2007.
3. McCormack, A.; Fortnum, H. Why do people fitted with hearing aids not wear them? *Int. J. Audiol.* **2013**, *52*, 360–368. [CrossRef] [PubMed]
4. Hamanishi, S.; Koike, T.; Matsuki, H.; Wada, H. A new electromagnetic hearing aid using lightweight coils to vibrate the ossicles. *IEEE Trans. Magn.* **2004**, *40*, 3387–3393. [CrossRef]
5. Hong, E.-P.; Park, I.-Y.; Seong, K.-W.; Cho, J.-H. Evaluation of an implantable piezoelectric floating mass transducer for sensorineural hearing loss. *Mechatronics* **2009**, *19*, 965–971. [CrossRef]
6. Liu, H.; Cheng, J.; Yang, J.; Rao, Z.; Cheng, G.; Yang, S.; Huang, X.; Wang, M. Concept and evaluation of a new piezoelectric transducer for an implantable middle ear hearing device. *Sensors (Basel)* **2017**, *17*, 2515. [CrossRef] [PubMed]
7. Song, Y.-L.; Jian, J.-T.; Chen, W.-Z.; Shih, C.-H.; Chou, Y.-F.; Liu, T.-C.; Lee, C.-F. The development of a non-surgical direct drive hearing device with a wireless actuator coupled to the tympanic membrane. *Appl. Acoust.* **2013**, *74*, 1511–1518. [CrossRef]
8. Koch, M.; Essinger, T.M.; Bornitz, M.; Zahnert, T. Examination of a mechanical amplifier in the incudostapedial joint gap: Fem simulation and physical model. *Sensors (Basel)* **2014**, *14*, 14356–14374. [CrossRef]
9. Colletti, V.; Soli, S.D.; Carner, M.; Colletti, L. Treatment of mixed hearing losses via implantation of a vibratory transducer on the round window. *Int. J. Audiol.* **2006**, *45*, 600–608. [CrossRef]
10. Zhao, S.; Gong, S.; Han, D.; Zhang, H.; Ma, X.; Li, Y.; Chen, X.; Ren, R.; Li, Y. Round window application of an active middle ear implant (amei) system in congenital oval window atresia. *Acta Otolaryngol.* **2016**, *136*, 23–33. [CrossRef]
11. Colletti, L.; Mandala, M.; Colletti, V. Long-term outcome of round window vibrant soundbridge implantation in extensive ossicular chain defects. *Otolaryngol. Head Neck Surg.* **2013**, *149*, 134–141. [CrossRef]
12. Nakajima, H.H.; Dong, W.; Olson, E.S.; Rosowski, J.J.; Ravicz, M.E.; Merchant, S.N. Evaluation of round window stimulation using the floating mass transducer by intracochlear sound pressure measurements in human temporal bones. *Otol. Neurotol.* **2010**, *31*, 506–511. [CrossRef] [PubMed]
13. Liu, H.; Xu, D.; Yang, J.; Yang, S.; Cheng, G.; Huang, X. Analysis of the influence of the transducer and its coupling layer on round window stimulation. *Acta Bioeng. Biomech.* **2017**, *19*, 103–111. [PubMed]
14. Schraven, S.P.; Hirt, B.; Gummer, A.W.; Zenner, H.P.; Dalhoff, E. Controlled round-window stimulation in human temporal bones yielding reproducible and functionally relevant stapedial responses. *Hear Res.* **2011**, *282*, 272–282. [CrossRef] [PubMed]
15. Maier, H.; Salcher, R.; Schwab, B.; Lenarz, T. The effect of static force on round window stimulation with the direct acoustic cochlea stimulator. *Hear Res.* **2013**, *301*, 115–124. [CrossRef] [PubMed]
16. Muller, M.; Salcher, R.; Lenarz, T.; Maier, H. The hannover coupler: Controlled static prestress in round window stimulation with the floating mass transducer. *Otol. Neurotol.* **2017**, *38*, 1186–1192. [CrossRef] [PubMed]
17. Nakajima, H.H.; Merchant, S.N.; Rosowski, J.J. Performance considerations of prosthetic actuators for round-window stimulation. *Hear. Res.* **2009**, *263*, 114–119. [CrossRef] [PubMed]
18. Salcher, R.; Schwab, B.; Lenarz, T.; Maier, H. Round window stimulation with the floating mass transducer at constant pretension. *Hear. Res.* **2014**, *314*, 1–9. [CrossRef]
19. Shin, D.H.; Cho, J.H. Piezoelectric actuator with frequency characteristics for a middle-ear implant. *Sensors (Basel)* **2018**, *18*, 1694. [CrossRef]
20. Arnold, A.; Ccandreia, S. Factors improving the vibration transfer of the floating mass transducer at the round window. *Otol. Neurotol.* **2010**, *31*, 122–128. [CrossRef]
21. Homma, K.; Du, Y.; Shimizu, Y.; Puria, S. Ossicular resonance modes of the human middle ear for bone and air conduction. *J. Acoust. Soc. Am.* **2009**, *125*, 968–979. [CrossRef]

22. Lerch, R. Simulation of piezoelectric devices by two- and three-dimensional finite elements. *IEEE Trans. Ultrason. Ferroelectr. Frequency Control* **1990**, *37*, 233–247. [CrossRef] [PubMed]
23. Schraven, S.P.; Hirt, B.; Goll, E.; Heyd, A.; Gummer, A.W.; Zenner, H.P.; Dalhoff, E. Conditions for highly efficient and reproducible round-window stimulation in humans. *Audiol. Neurootol.* **2012**, *17*, 133–138. [CrossRef]
24. Gan, R.Z.; Cheng, T.; Dai, C.; Yang, F.; Wood, M.W. Finite element modeling of sound transmission with perforations of tympanic membrane. *J. Acoust. Soc. Am.* **2009**, *126*, 243–253. [CrossRef]
25. Puria, S.; Peake, W.T.; Rosowski, J.J. Sound-pressure measurements in the cochlear vestibule of human-cadaver ears. *J. Acoust. Soc. Am.* **1997**, *101*, 2754–2770. [CrossRef] [PubMed]
26. Aibara, R.; Welsh, J.T.; Puria, S.; Goode, R.L. Human middle-ear sound transfer function and cochlear input impedance. *Hear. Res.* **2001**, *152*, 100–109. [CrossRef]
27. Stieger, C.; Rosowski, J.J.; Nakajima, H.H. Comparison of forward (ear-canal) and reverse (round-window) sound stimulation of the cochlea. *Hear. Res.* **2013**, *301*, 105–114. [CrossRef] [PubMed]
28. Stenfelt, S.; Puria, S.; Hato, N.; Goode, R.L. Basilar membrane and osseous spiral lamina motion in human cadavers with air and bone conduction stimuli. *Hear. Res.* **2003**, *181*, 131–143. [CrossRef]
29. Gundersen, T.; Skarstein, O.; Sikkeland, T. A study of the vibration of the basilar membrane in human temporal bone preparations by the use of the mossbauer effect. *Acta Otolaryngol.* **1978**, *86*, 225–232. [CrossRef]
30. Homma, K.; Shimizu, Y.; Kim, N.; Du, Y.; Puria, S. Effects of ear-canal pressurization on middle-ear bone- and air-conduction responses. *Hear. Res.* **2010**, *263*, 204–215. [CrossRef]
31. Areias, B.; Parente, M.P.L.; Santos, C.; Gentil, F.; Natal Jorge, R.M. The human otitis media with effusion: A numerical-based study. *Comput. Methods Biomech. Biomed. Eng.* **2017**, *20*, 958–966. [CrossRef]
32. Liu, H.; Rao, Z.; Huang, X.; Cheng, G.; Tian, J.; Ta, N. An incus-body driving type piezoelectric middle ear implant design and evaluation in 3d computational model and temporal bone. *Sci. World J.* **2014**, *2014*, 121624. [CrossRef] [PubMed]
33. Gan, R.Z.; Dai, C.; Wang, X.; Nakmali, D.; Wood, M.W. A totally implantable hearing system–design and function characterization in 3d computational model and temporal bones. *Hear. Res.* **2010**, *263*, 138–144. [CrossRef] [PubMed]
34. Lee, C.-F.; Chen, J.-H.; Chou, Y.-F.; Liu, T.-C. The optimal magnetic force for a novel actuator coupled to the tympanic membrane: A finite element analysis. *Biomed. Eng. Appl. Basis Commun.* **2007**, *19*, 171–177. [CrossRef]
35. Atturo, F.; Barbara, M.; Rask-Andersen, H. Is the human round window really round? An anatomic study with surgical implications. *Otol. Neurotol.* **2014**, *35*, 1354–1360. [CrossRef]
36. Ishii, T.; Takayama, M.; Takahashi, Y. Mechanical properties of human round window, basilar and reissner's membranes. *Acta Otolaryngol. Suppl.* **1995**, *519*, 78–82. [CrossRef] [PubMed]
37. Schwab, B.; Grigoleit, S.; Teschner, M. Do we really need a coupler for the round window application of an amei? *Otol. Neurotol.* **2013**, *34*, 1181–1185. [CrossRef] [PubMed]
38. ASTM F2504-05. *Standard Practice for Describing System Output of Implantable Middle Ear Hearing Devices*; ASTM International: West Conshohocken, PA, USA, 2014.
39. Arnold, A.; Kompis, M.; Candreia, C.; Pfiffner, F.; Häusler, R.; Stieger, C. The floating mass transducer at the round window: Direct transmission or bone conduction? *Hear. Res.* **2010**, *263*, 120–127. [CrossRef]
40. Bornitz, M.; Hardtke, H.J.; Zahnert, T. Evaluation of implantable actuators by means of a middle ear simulation model. *Hear. Res.* **2010**, *263*, 145–151. [CrossRef] [PubMed]
41. Maurer, J.; Savvas, E. The esteem system: A totally implantable hearing device. *Adv. Otorhinolaryngol.* **2010**, *69*, 59–71.
42. Wang, Z.; Mills, R.; Luo, H.; Zheng, X.; Hou, W.; Wang, L.; Brown, S.I.; Cuschieri, A. A micropower miniature piezoelectric actuator for implantable middle ear hearing device. *IEEE Trans. Biomed. Eng.* **2011**, *58*, 452–458. [CrossRef]

 *micromachines*

Article

Development of ECG Monitoring System and Implantable Device with Wireless Charging

Jae-Ho Lee [1] and Dong-Wook Seo [2,]*

[1]   Electronic and Telecommunications Research Institute, Daegu 42994, Korea; jhlee1229@etri.re.kr
[2]   Department of Radio Communication Engineering, Korea Maritime and Ocean University,
    Busan 49112, Korea
*   Correspondence: dwseo@kmou.ac.kr; Tel.: +82-51-410-4427

Received: 20 November 2018; Accepted: 4 January 2019; Published: 8 January 2019

Abstract: We developed an implantable electrocardiogram (ECG) monitoring system and demonstrated its performance through an in vivo test. In the system, the implantable device senses not only the ECG signal of the animal but also the voltage level of the secondary cell and temperature inside the implantable device, and users can check the transmitted information through a PC program or a mobile application. The adoption of wireless charging technology eliminates the use of a lead wire and repetitive surgery to replace the implantable device. The proposed wireless charging technology demonstrated experimentally a wireless power transfer efficiency of approximately 30%. To minimize the size of the implantable device, the antenna and coil were integrated into a size of 34 mm × 14 mm. Communication between the implantable device and the basestation can reach up to 2.4 m when the implantable device is inserted into a porcine skin sample.

Keywords: electrocardiogram (ECG); implantable device; MedRadio; wireless charging; wireless power transfer

---

1. Introduction

As the fourth industrial revolution has accelerated, Internet of Things (IoT) technologies have spread deeply into human lives. In the early stage, IoT technologies were mainly applied to home appliances as additional convenience features. Recently, the application of IoT technology has extended beyond the simple control of objects to the sensing of human bio-signals.

As bio-signals are generally very weak, biosensors should have contact with the human body. Electrocardiogram (ECG) sensors used to monitor arrhythmia have been developed primarily as patch types [1,2]. Recently, many of the emerging sensors such as electric, optic, and chemical sensors have been introduced to sense various physiological signals on the epidemies [3]. However, existing patch-type ECG sensors should be attached to the human body, causing serious inconvenience to people in need of service. For this reason, many biosensors have evolved into implant or remote sensing devices. Remote sensing devices have been mainly developed using radar technologies, which sense changes in periodic motions such as heart rate and respiratory function [4,5]. On the other hand, the implantable devices directly sense the bio-signals inside the human body and transfer the sensed signals to a device outside the body. The implantable devices are mainly powered by the primary cells. However, the primary cells have a limited life, and thus additional surgery is required to replace dead cells on a regular basis. To relieve these problems, wireless power transfer (WPT) technology has been applied to devices that are implanted into shallow organs through the skin [6–12]. The functions of the implantable devices can be largely classified into three categories: Monitoring, treatment, and auxiliary. First, monitoring devices sense and observe the patient's vital signals [5,6]; a typical example of such a device is the loop recorder which records heartbeats. Second, examples of therapeutic implantable devices include pacemakers and dorsal stimulators [7–9]. These devices directly drive an electrical stimulation to organs such as the heart or nerve. Finally, auxiliary implantable devices include cochlear implants and retinal implants. They change external sounds or lights to electrical signals and deliver the changed signals to the auditory or optic nerves [11,12]. Therapeutic or auxiliary implantable devices are limited to people suffering from special diseases or handicaps, whereas monitoring implantable devices can be applied to an ordinary person. In particular, implant monitoring systems can be effective for people living in areas where immediate medical services are not available.

Most of the state-of-art implantable devices work only when powered wirelessly from out-body [13]. Previous ECG implantable devices also work in a similar way [5,6]. For small animals, there is no problem even if the implantable device operates only during wireless power transmission. However, for patients with a chronic disease, inadequate mobility, or arrhythmia, it is necessary to continuously monitor vital signs 24/7. In this case, it is virtually impossible to continuously supply power from the outside. To solve this problem, in this paper, we adopt WPT technology to wirelessly charge the secondary cell in an implantable device. This solution can continue to operate the implantable device until the secondary cell is discharged. Additionally, we demonstrate a system for monitoring an ECG signal, one of the most basic of human bio-signals, from the subcutaneous fat of the human body.

2. Prototype and Demonstration of ECG Monitoring System

Figure 1 shows a diagram of the developed prototype used to monitor ECG signals. An ECG sensor is implanted in the subcutaneous fat to sense the ECG signals and transfer the sensed signal to the basestation through the medical device radio communication service (MedRadio) band. The basestation collects the ECG signals and transfers power to the implantable device using WPT technology when the battery voltage level of the implantable device is low. A personal terminal is connected to the basestation through Bluetooth, and the user checks and monitors the status of the person with the implantable device.

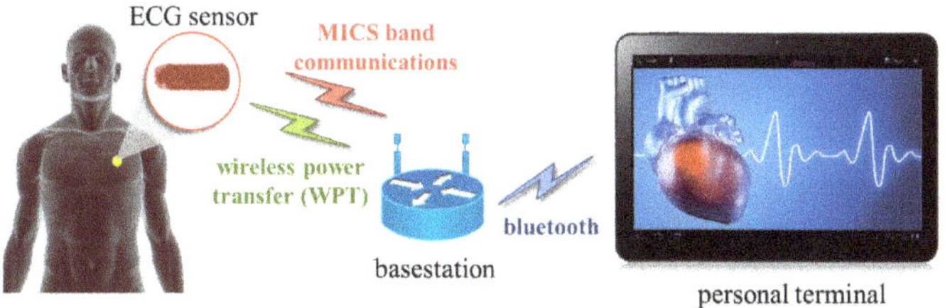

Figure 1. Electrocardiogram (ECG) monitoring system diagram.

Figure 2 shows the developed prototype modules. The dimensions of the implantable device and the basestation are 19.4 mm × 55.4 mm × 9 mm and 71 mm × 77 mm × 22 mm, respectively. In the personal terminal, the developed monitoring application shows the ECG signals of the user and the temperature inside the implantable device, as well as the battery voltage level of the implantable device.

As shown in Figure 3, the prototype was implanted into a 40 kg Hanford (HAN) miniature pig to demonstrate its performance. Because the implantable device is small, only a small incision was required. The implantable device was implanted in the subcutaneous fat layer of the chest.

After the device was implanted, the collected ECG signals and temperature were displayed on the monitoring program in real time, as shown in Figure 4a. In addition, power was wirelessly transferred to the implantable device from an external transmitting coil just above the skin, as shown in Figure 4b. Figure 5 shows a photograph of the monitor program displaying three types of information. As the ECG signal of the pigs differs from that of humans, the ECG signal in Figure 5 looks unusual, but works properly with periodic signals. The program displays the sensed temperature of 22.5 °C, while the body temperature of a typical pig is approximately 40 °C [14]. This discrepancy is caused by not measuring the contact body temperature but measuring the temperature inside the implantable device. Finally, we recharged the secondary cell wirelessly for 1 h after fully discharging it to check the wireless charging function. The implantable device operated normally, and the charged battery had a voltage of approximately 3 V, as shown in Figure 5.

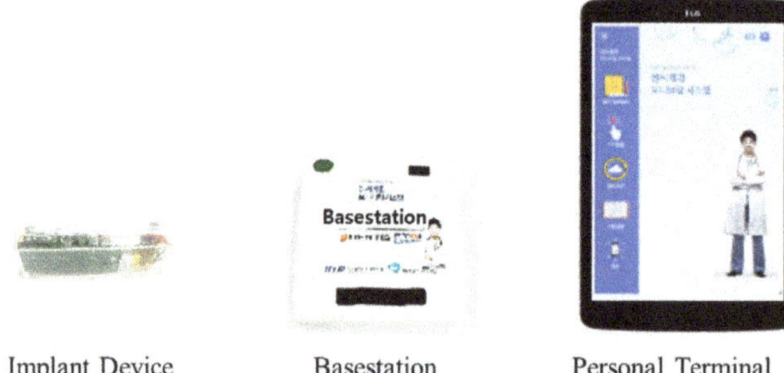

Figure 2. Developed prototype modules of the ECG monitoring system.

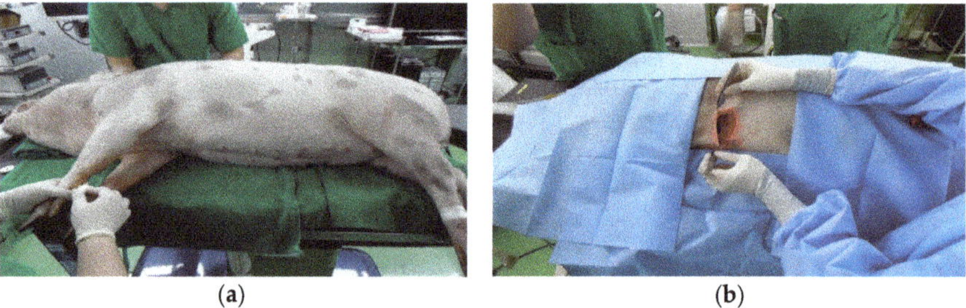

Figure 3. Animal experiment: (**a**) Test animal and (**b**) photograph of implanting the device into subcutaneous fat.

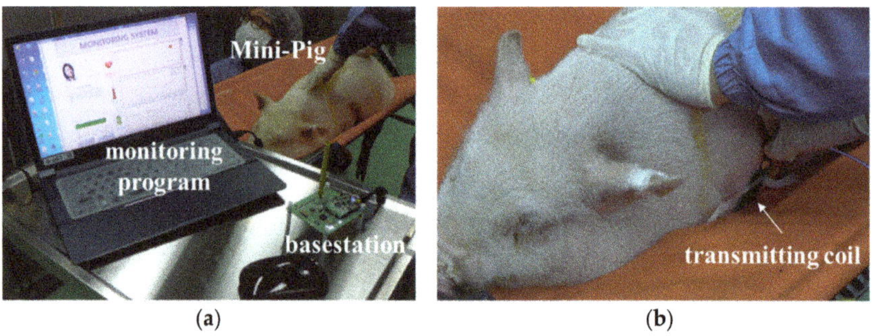

Figure 4. Animal experiment: (**a**) Experiment setup and (**b**) wireless charging.

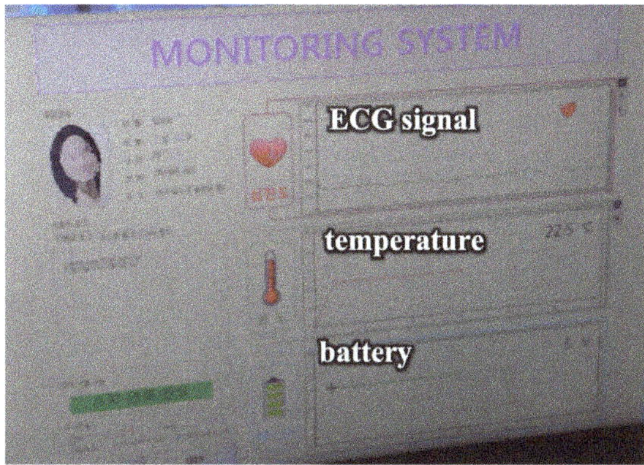

Figure 5. Sensed bio-signals and battery voltage level on the monitoring program.

3. Implantable Device

Figure 6 shows the developed implantable device consisting of a system on a board (SoB), an antenna and a coil board, a secondary cell, and a package.

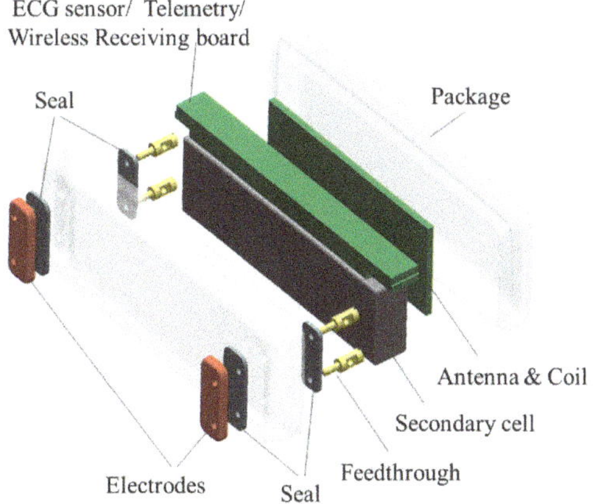

Figure 6. Configuration of the implantable device.

For long-term operation of the device with a single charge, a secondary cell with a large capacity is required. In general, the capacity of the secondary cell increases with its volume, and thus a cell with a large volume is required for a long operation period, which undesirably increases the dimensions of the implantable device. With the maximally optimized integration of an electric circuit on the board, the ECG sensor, telemetry and control unit, and WPT circuits, the implantable device was implemented in a single board of 50 mm × 5 mm, and the SoB was placed on the side of the secondary cell. In contrast, the antenna for the telemetry and the coil for the wireless charging should be large so as to secure a large effective area in the electromagnetic receiving ends. To this end, the antenna and coil were integrated and placed on the front side of the secondary cell.

Figure 7 shows a photograph of the developed SoB upon which the ECG sensor, temperature sensor, main control unit (MCU), radio-frequency (RF) transceiver for telemetry, and WPT receiving unit are integrated. In this section, we describe these units of the implantable device in detail.

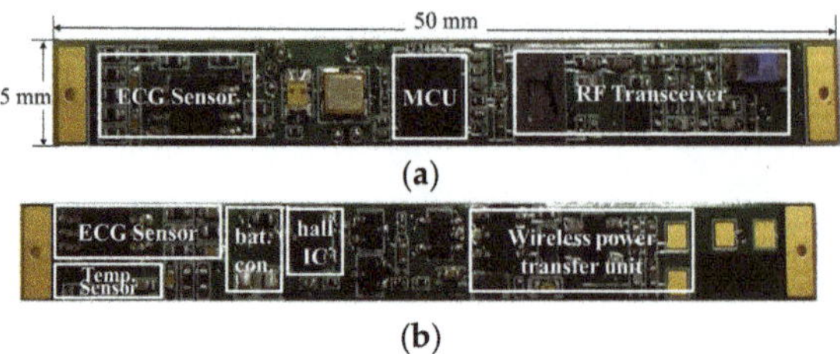

(a)

(b)

Figure 7. The developed system on a board (SoB) for ECG sensing, telemetry, and wireless power transfer (WPT) receiving: (**a**) Front and (**b**) back sides.

3.1. Sensor Unit

To capture the electric potential generated by the heart, the two electrodes must be on different equal-potential lines. This means that the distance between the two electrodes should be several centimeters or more [15]. Based on the measurement of the ECG signal according to the distance between the two electrodes, a minimum distance of 35 mm is required to obtain a stable ECG signal. Thus, the pads for connecting two electrodes were placed on both sides of the SoB, as shown in Figure 7.

The structure of the ECG sensor is shown in Figure 8. Because the heart-generated electric potential is low, within the range of 1–5 mV, the instrumentation amplifier was implemented using a TI INA333 (Texas Instruments, Austin, TX, USA) to amplify the sensed differential signals. Additionally, the high- and low-pass filters were cascaded, and the notch filter eliminated the power line noise of 60 Hz. The output of the ECG sensor entered the analog digital converter (ADC) of the MCU.

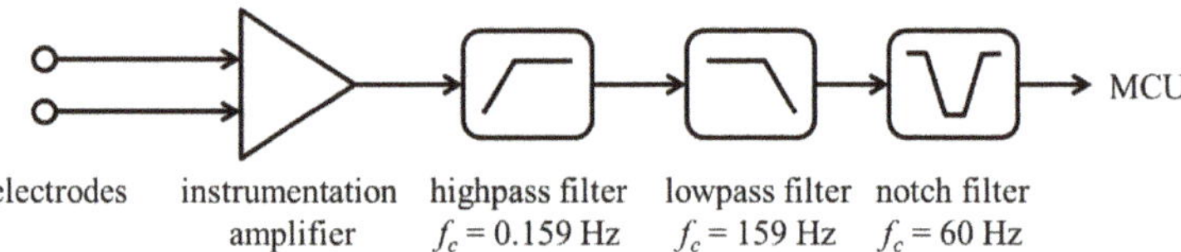

Figure 8. Block diagram of the ECG sensor.

At the development stage, we received the ECG signal from the sensor after attaching commercial ECG patches on the wrists or chest, as shown in Figure 9. Based on the sensed ECG signal, we tuned the cutoff frequencies of the high- and low-pass filters. The ECG signal is so weak that it is easily contaminated from a 60 Hz noise generated by surrounding power lines. Actually, a noise of 60 Hz was observed when the battery as well as power supply were powered to the sensor. Therefore, a 60 Hz notch filter was applied to the SoB.

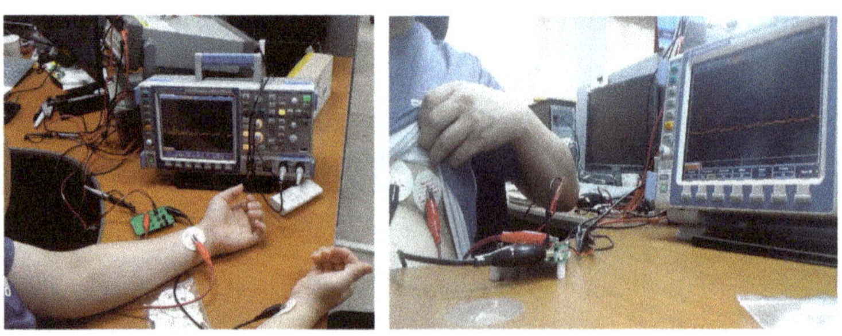

Figure 9. Testing the ECG signals from the wrists and chest.

Figure 10 describes the experiment setup used to measure the power consumption of the ECG sensor module. The ECG simulator provided a virtual ECG signal, and only the parts related to the ECG sensor were turned on. For the power supply, the current provided at a supply voltage of 3 V was 0.3–0.4 mA during an active state. A Murata NCP series part was used to measure the temperature inside the implantable device, and a voltage-distributed circuit, configured for a series of resistances, was used to measure the voltage level of the secondary cell. According to the part's datasheet, the operating temperature range was from -40 to 125 °C, but the temperature sensor was calibrated at 5-degree intervals from 20 to 40 °C, and we confirmed that the temperature sensor normally senses the right value under the circuit board.

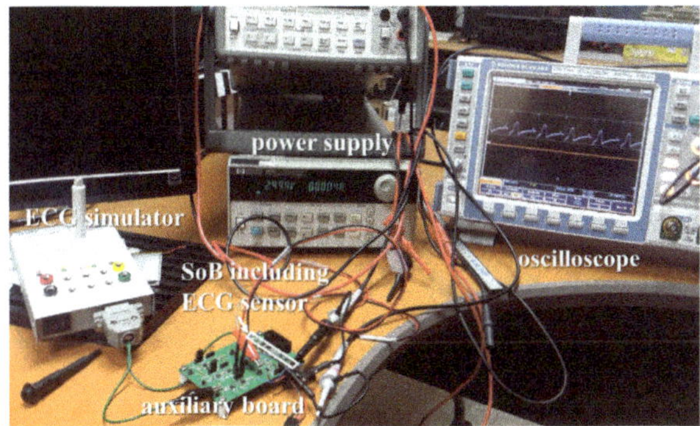

Figure 10. Power consumption test of the ECG sensor.

3.2. Telemetry and Control Unit

Figure 11 shows the block diagram of the SoB. The output data of the sensor units, such as the ECG, temperature, and battery voltage, were input to the ADC ports of the MCU and converted into digital data. The MCU processed and stored the imported sensing data and controled a RF transceiver as well as the sensors using the serial peripheral interface bus (SPI). The used RF transceiver and MCU were the Microsemi's ZL70102 and the TI's MSP430, respectively.

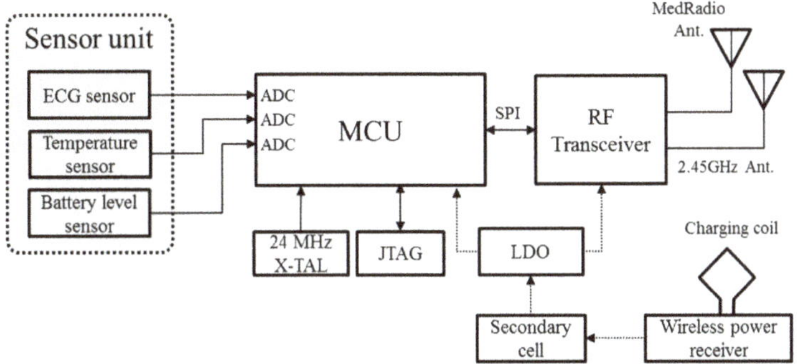

Figure 11. Block diagram of the SoB.

Next, the converted digital data was transmitted to the basestation through the RF link. Of the various frequency bands, the MedRadio has been allocated in the 401–406 MHz for data transmission of an implanted medical device by the Federal Communication Commission (FCC) [16]. The ZL70102 transceiver operates in the MedRadio for telemetry and 2.45 GHz industrial scientific medical (ISM) band for waking up the sleep mode of the transceiver, respectively. Therefore, two antennas for

MedRadio and 2.45 GHz ISM band were installed, as shown in Figure 11; a chip antenna for the 2.45 GHz frequency band and a printed antenna for the MedRadio band. The printed antenna is described in detail in the Section 3.4.

Since the implantable device operates with limited battery capacity, low power consumption is one of the most important considerations. To measure the power consumption, we measured the DC current supplied by the secondary cell in the active and sleep modes, as shown in Figure 12. The DC rated supply voltage of the secondary cell was 3.0 V, so the estimated power consumption was 18.82 mW and 0.09 mW in the active and sleep modes, respectively.

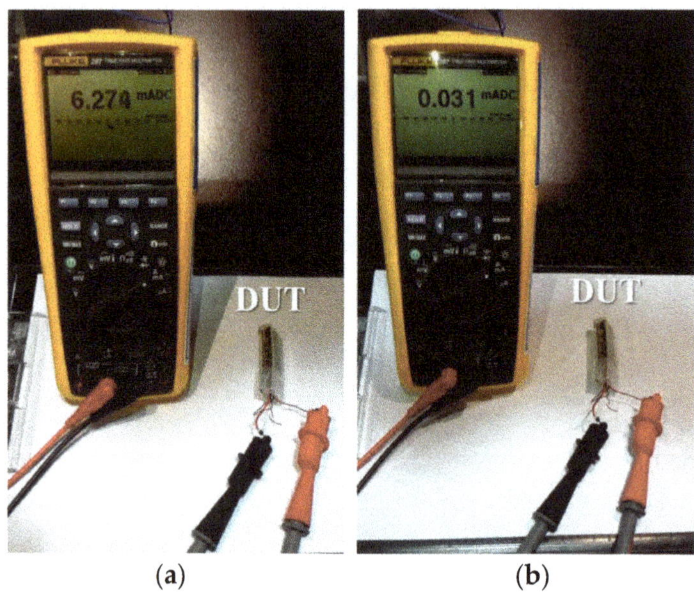

(a)   (b)

Figure 12. Experiments for measuring the power consumption (**a**) in the active mode and (**b**) in sleep mode.

The maximum communication range is another figure of merit of an implantable device. When the implantable device is on the desk, the measured maximum range was 4.2 m. However, since the implantable device should ultimately work in the human body, the communication range was measured with the implantable device located in the human body mimicking material called a phantom. As shown in Figure 13, we inserted the implantable device in the phantom and a slice of pork and measured the maximum communication range by varying the distance between the device and basestation. The experimental results confirm that the maximum communication range of the implantable device with the basestation was 2.4 m for the phantom and 2.8 m for the pork environments.

In addition, we also measured the maximum data rate from the implantable device to the basestation. Figure 14 shows that the implantable device can transmit all data, including the sensing and control signals, to the basestation at a data rate of 127 kbps. In this case, the 12-bit ADC of the MCU was operated with sampling rate of 100 Hz.

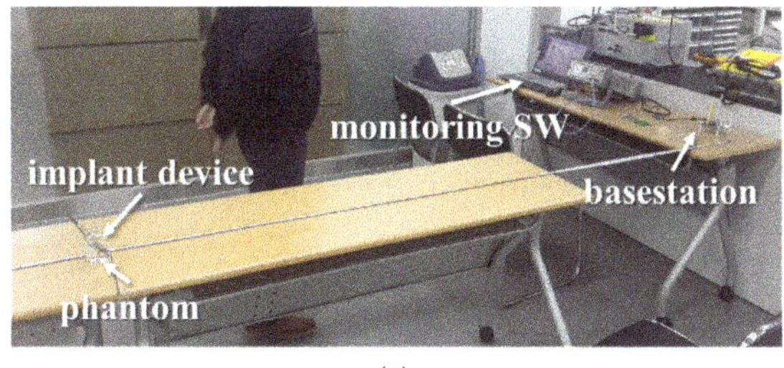

(**a**)

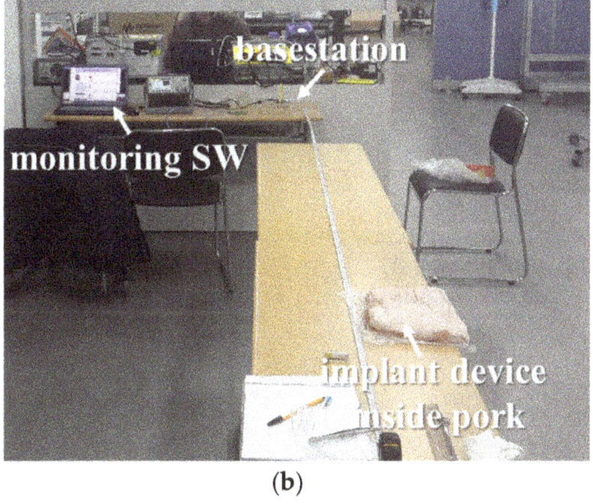

(**b**)

Figure 13. Experiment setup used to measure the maximum communication range when the implantable device is placed inside the (**a**) phantom and (**b**) pork.

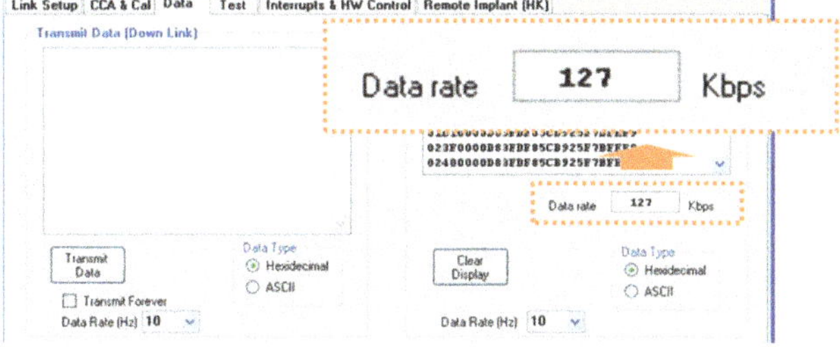

Figure 14. Data rate test result.

3.3. Wireless Power Transfer (WPT) Unit

The 13.56 MHz power signal from the receiving coil of the implantable device was converted into DC power by the WPT receiving unit, and the converted DC power was used to charge the secondary cell.

Figure 15 shows a block diagram of the WPT receiving unit. The receiving and transmitting coils were designed to have a high Q-factor at an operating frequency of 13.56 MHz. The matching network with an L-section structure has two functions: One for generating resonance and the other for transferring the coil impedance to the input impedance of the full bridge rectifier.

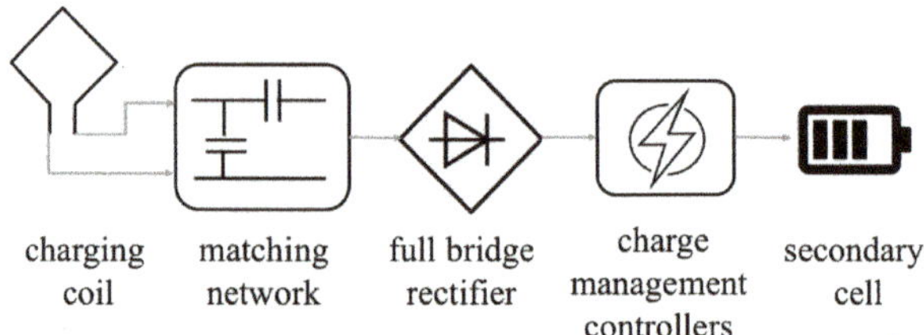

Figure 15. Block diagram of the WPT receiving unit.

The maximum distance between the transmitting and receiving coils was about 20 mm because the implantable device was inserted into the subcutaneous fat and the transmitting coil was located on the skin. For this reason, we designed the matching network to achieve the maximum power transfer efficiency (PTE) at a distance of 20 mm. A full bridge rectifier converting AC to DC power was implemented using an Avago HSMS2828. The linear charge management controller, a Microchip MCP73831, effectively charges the secondary cell depending on the battery status, and the user can check the charge status through the LED which emits red light only when charging is in progress. The LED light can be used only for in vitro experiment, therefore, when the device is implanted in a human body, the user can check whether the device is being charged by receiving the battery voltage.

Biological tissue does not have magnetic properties [17], and thus has little effect on the magnetic field and PTE. Nevertheless, we tested the wireless charging using pork, as shown in Figure 16. The lit LED indicates that the wireless charging function is working properly. To estimate the accurate charging power and PTE, we measured the charging current and voltage, which are the output of the linear charge management controller, as shown in Figure 17. At a distance of 20 mm, the charging power was estimated to be 75.66 mW (19.767 mA × 3.8257 V). The total PTE was also estimated to be 30.26% based on the 250 mW of maximum available power of the function generator. There is a typical trade-off between the PTE and the size of the receiving coil, and thus the larger the size of the receiving coil, the higher the PTE.

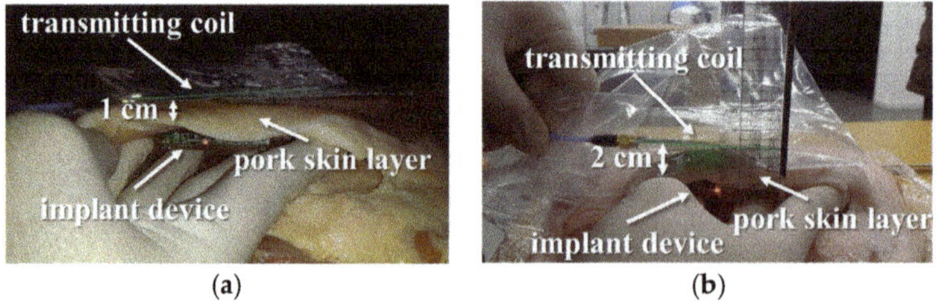

Figure 16. Wireless charging test in pork at distances of (**a**) 1 and (**b**) 2 cm.

3.4. Antenna for MedRadio and Coil for Wireless Charging

For the implantable device to communicate with the basestation and wirelessly charge, the antenna and receiving coil should face outward. The larger the size of the units, the higher the gain and the quality factor (Q-factor). To design a compact antenna and coil, we integrated them as described in Reference [18]. As the key feature of the design, a high-permeability sheet, such as ferrite, was introduced to the dielectric substrate to minimize the antenna and coil size and enhance the operating bandwidth of the antenna.

The integrated coil and antenna were designed as shown in Figure 18; the antenna is inside the integrated surface, and the coil surrounds the outside of the antenna. Compared with the structure in Reference [18], the size of this version increased from 30 mm × 5 mm to 34 mm × 14 mm, making full use of the area of the secondary cell. For the coil, the number of turns was reduced from 4 to 2, and the line width widened through the optimization process. The antenna has a meandering planar inverted-F antenna (PIFA) structure, the designed parameters for which are summarized in Table 1.

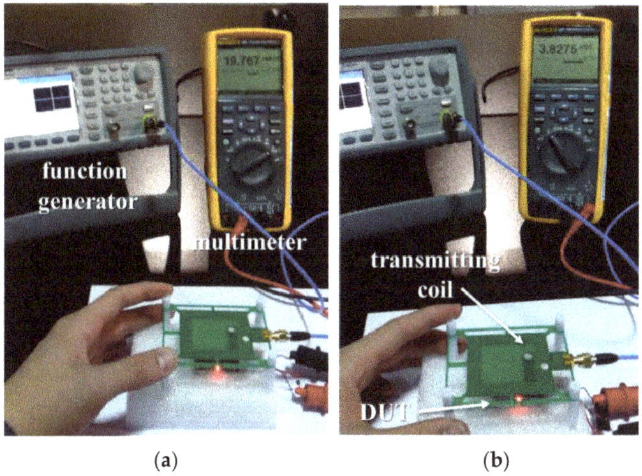

(a)   (b)

Figure 17. Measurement setup for (**a**) charging current and (**b**) voltage of the secondary cell.

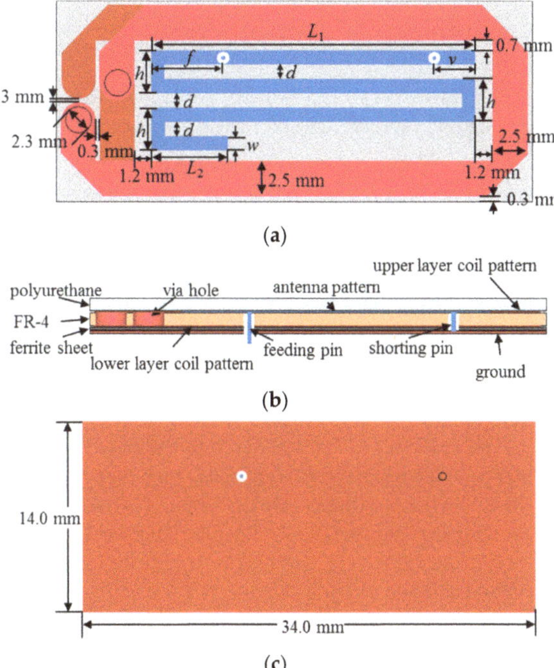

Figure 18. Geometry of the resonant coil and MedRadio antenna: (**a**) Top, (**b**) side, and (**c**) bottom.

Table 1. Parameters of the designed antenna.

Parameters	L_1	L_2	d	h	v	f
value (mm)	23.2	5.5	1	3	3	5.2

The characteristics of the antenna are described in detail in Reference [19]; however, the effects of the outer coil are not included. Using the phantom, we measured the return loss of the antenna with the coil, the simulation results of which are shown in Figure 19. As the simulation results indicate, the antenna has a broad bandwidth of 380 MHz with a −10 dB bandwidth criterion. However, the measured results show a bandwidth of 264 MHz, which is sufficiently wide to cover the MedRadio frequency band of 401–406 MHz.

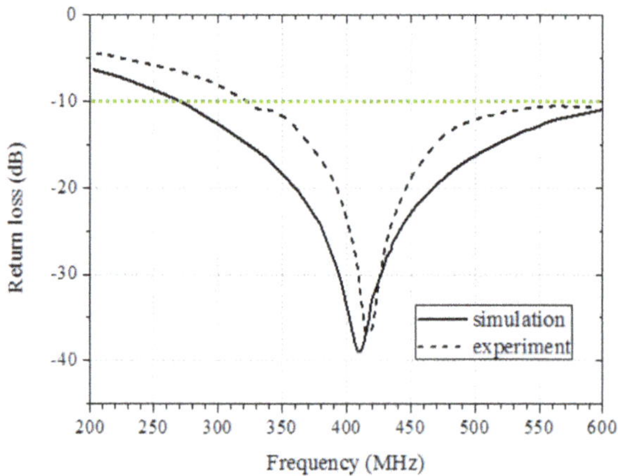

Figure 19. Reflection coefficient frequency responses with respect to frequency.

In the simulation, the antenna showed a gain of −30 dBi at 403 MHz. Based on the regulation that the equivalent isotropic radiated power (EIRP) of the MedRadio antenna should be limited to 25 μW (−46 dBm), the input power of the antenna should be less than −16 dBm. The input power can be controlled from −30 to −1 dBm using a RF transceiver.

For the implantable antenna, the maximum specific absorption rate (SAR) is also a significant consideration. The IEEE C95.1 standard restricts the spatial SAR peak to less than 1.6 W/kg (SAR1g, max ≤ 1.6 W/kg) as averaged over any 1 g of cubic-shaped tissue [20]. According to the EM simulation, the antenna has a 1-g SAR of 39.4 W/kg (peak SAR value) when 1 W of power is delivered to the antenna, which indicates that the delivered power should be less than 40 mW. In our system, the maximum power from the RF transceiver is 1 mW, and therefore the implantable device complies with the above-mentioned SAR restriction.

The system parameters of the coil are summarized in Table 2. The Q-factor of the coil is about twice that of the previous coil described in Reference [18]. In addition, as a result of the increased size of the receiving coil, the size of the transmitting coil is also increased, and the transmitting coil has a larger Q-factor than that of the transmitting coil in Reference [18]. As a result, the efficiency of the link is increased by approximately three times. The link efficiency shown in Table 2 is defined as the efficiency between two coils, which is the efficiency excluding the influence of the low dropout (LDO) regulator in the PTE mentioned above.

Table 2. Parameters and performance of the receiving coil.

Parameters	L	R	Q-Factor	Link Eff.
value	0.18 μH	0.316 Ω	48.48	0.61

3.5. Thermal Analysis

Implantable sensors are known to potentially induce inflammation of body tissues if a temperature change of more than 1 °C occurs. For this reason, the average temperature change of an implantable device should be minimized. As shown in Figure 20, we took snapshots of the temperature distribution on the SoB using an infrared (IR) camera and performed a thermal analysis. First, we distinguished the major hot spots and the parts causing hot spot, and then placed them apart. Finally, the output power of the RF transceiver was adjusted to satisfy a temperature change of 1 °C or less.

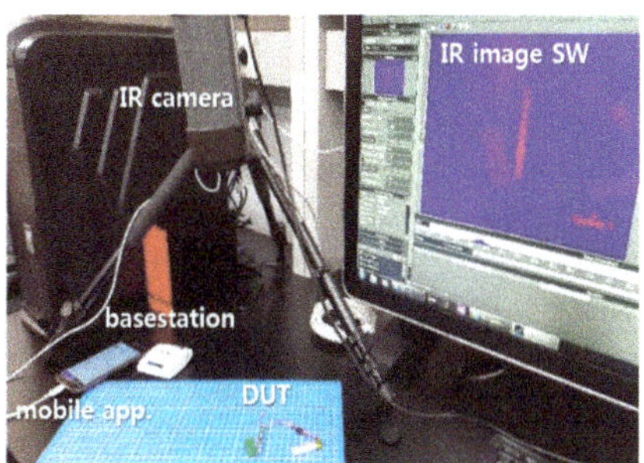

Figure 20. Measurement setup for measuring the temperature distribution on the SoB.

After applying the thermal analysis results to the SoB, we observed the temperature distribution for 30 min during which the monitoring system was operating, as shown in Figure 21. The changes of average and maximum temperatures over time are shown in Figure 21b. Compared with the previous version, the maximum temperature decreased by 0.5 °C at the RF transceiver, and the change of the average temperature also decreased from 2.2 °C to 0.9 °C. Consequently, the overall temperature change in this case was less than 1 °C.

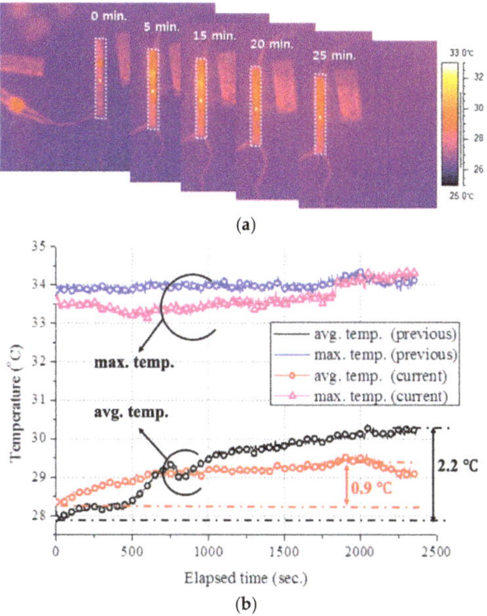

Figure 21. Measured thermal characteristics of the revised SoB using a thermal analysis: (**a**) Snapshots of the thermal distribution and (**b**) temperature changes.

3.6. Battery and Package

A lithium polymer battery with a capacity of 165 mAh was used to supply the operating power. As mentioned before, the implantable device consumes 6.27 mA and 0.03 mA of current during an active and idle state, respectively. Therefore, a capacity of 165 mAh is computationally sufficient to operate the implantable device all day long without idle. In actual operation, the implantable device does not continuously detect the signal for a 24 h-period, and alternates between idle and active modes depending on the user. We set the active mode to 5 s and the idle mode to 160 s. Thus, the operating time of the implantable devicewith a single charge can be calculated from the following equation, and the calculated duration time was 31.2 days.

$$165\,\text{mAh} = \text{Days} \times 24\,\text{h} \times \frac{6.274\,\text{mA} \times 5\,\text{s} + 0.031\,\text{mA} \times 160\,\text{s}}{5\,\text{s} + 160\,\text{s}} \tag{1}$$

However, a digital omnipolar magnetic switch (MDT MMS2X1H) connected to the LDO regulator enables the user to forcibly turn the implantable device on and off using an external magnet if necessary. This function can prevent unnecessary power consumption by turning off the power when the sensor is not used before or after putting it into the human body.

Biocompatibility is the most stringent precondition for implantable devices; the implantable device is safe to insert into the body and can operate in vivo. The biocompatibility is mainly affected by the packaging material. The most commonly used materials for biocompatible materials are cobalt-chromium, iridium, titanium, platinum, nitinol, specific glass, polydimethylsiloxane (PDMS), polyurethane, polymethylmethacrylate (PMMA), and polyimide [21]. Polyurethane with excellent electrical insulation characteristics was chosen as the packaging material. The package was sealed with ultrasonic bonding. On the other hand, electrodes are needed to measure the electric potential generated by the heart. Generally, wet electrodes such as Ag/AgCl electrodes are not suitable for long-term cardiac monitoring due to skin irritation and allergic reactions [22]. Among various dry electrodes, titanium exhibit excellent stability in vivo, so we used the electrodes fabricated using titanium.

4. Basestation Module

The basestation receives the bio- and control signals from the implantable device through the MedRadio band, and retransmits the signals to the user terminal via Bluetooth 4.0 low energy (BLE), as shown in Figure 22. Except for Bluetooth, most of the blocks in the basestation are similar to the SoB in the implantable device. The basestation includes LEDs indicating the charging state and connection states between the implantable device and the user terminal.

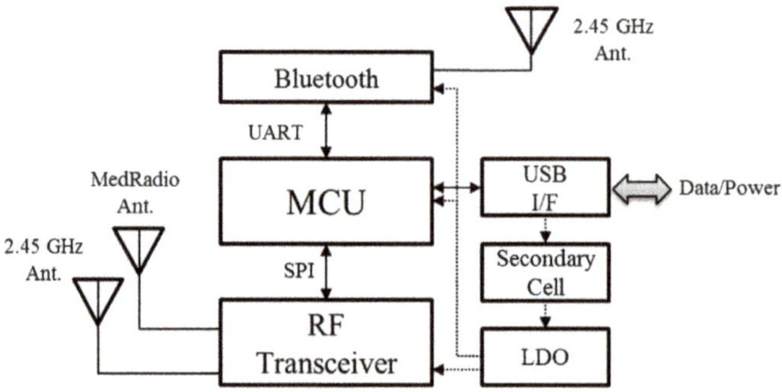

Figure 22. Block diagram of the basestation.

A commercial helical antenna is used for the MedRadio band, and chip antennas are used for the 2.45 GHz wake-up link and Bluetooth.

The power unit consists of a 1500 mhA lithium-polymer cell and LDOs, and users can control the power with a slide switch. A mini USB is used to charge the secondary cell through a BQ24080 charger and additionally transfer data to an external device via wire, and the charging status LED indicates whether the battery is being charged.

5. Conclusions

We developed an implantable ECG monitoring system and demonstrated its performance through an in vivo test using an animal. The ECG signal and temperature of the animal, as well as the voltage level of the secondary cell, were received from the implantable device and were confirmed using a PC program.

The implantable device with wireless charging technology does not require lead wires for the power supply, and can be charged even when discharged. The applied wireless charging technology shows a PTE of approximately 30%. A compact implantable device was developed by integrating a MedRadio antenna and a wireless charging coil. When implanted in pork, the developed device can communicate with the external basestation at up to a distance of 2.8 m. The implanted device consumes a current of 6.27 mA in an active state, and the adopted battery capacity is 165 mAh; therefore, it is sufficient to charge the implanted device once a month even if an ECG signal is continuously sensed. In addition, the system can be used to monitor cardiac arrhythmias and additionally manage experimental animals.

Author Contributions: J.-H.L. and D.-W.S. conceived and designed the experiments; J.-H.L. and D.-W.S. performed the experiments; J.-H.L. and D.-W.S. analyzed the data; J.-H.L. wrote the paper; D.-W.S. revised the paper.

Funding: This research was supported by the National Research Foundation of Korea (NRF) grant funded by the Korea government (MSIT) (No. NRF-2018R1C1B6003854) & Electronics and Telecommunications Research Institute (ETRI) grant funded by the Korean Government (18ZD1120, Development of ICT Convergence Technology for Daegu-GyeongBuk Regional Industry).

Conflicts of Interest: The authors declare no conflict of interest.

References

1. Yamamoto, D.; Nakata, S.; Kanao, K.; Arie, S. All-Printed, Planar-Type Multi-Functional Wearable Flexible Patch Integrated with Acceleration, Temperature, and ECG Sensors. In Proceedings of the 2017 IEEE International Conference Micro Electro Mechanical System (MEMS), Las Vegas, NV, USA, 22–26 January 2017; pp. 239–242.

2. Zheng, K.; Chen, S.; Zhu, L.; Zhao, J.; Guo, X. Large Area Solution Processed Poly (Dimethyl siloxane)-Based Thin Film Sensor Patch for Wearable Electrocardiogram Detection. *IEEE Electron Device Lett.* **2018**, *39*, 424–427. [CrossRef]

3. Heikenfel, J.; Jajack, A.; Rogers, J.; Gutruf, P.; Tian, L.; Pan, T.; Li, R.; Khine, M.; Kim, J.; Wang, J.; Kim, J. Wearable Sensors: Modalities, Challenges, and Prospects. *Lab Chip* **2019**, *18*, 217–248. [CrossRef] [PubMed]

4. Park, J.; Ham, J.-W.; Park, S.; Kim, D.-H.; Park, S.-J.; Kang, H.; Park, S.-O. Polyphase-Basis Discrete Cosine Transform for Real-Time Measurement of Heart Rate with CW Doppler Radar. *IEEE Trans. Microw. Theory Tech.* **2018**, *66*, 1644–1659. [CrossRef]

5. Tu, J.; Lin, J. Fast Acquisition of Heart Rate in Noncontact Vital Sign Radar Measurement Using Time-Window-Variation Technique. *IEEE Trans. Instrum. Meas.* **2016**, *65*, 112–122. [CrossRef]

6. Lee, J.-H. Human Implantable Arrhythmia Monitoring Sensor with Wireless Power and Data Transmission Technique. *Austin J. Biosens. Bioelectron.* **2015**, *1*, 1008.

7. Lee, J.-H. Miniaturized Human Insertable Cardiac Monitoring System with Wireless Power Transmission Technique. *J. Sens.* **2016**, *2016*, 5374574. [CrossRef]

8. Ho, J.S.; Yeh, A.J.; Neofytou, E.; Kim, S.; Tanabe, Y.; Patlolla, B.; Beygui, R.E.; Poon, A.S. Wireless Power Transfer to Deep-tissue Microimplants. *Proc. Natl. Acad. Sci. USA* **2014**, *111*, 7974–7979. [CrossRef] [PubMed]

9. Kim, S.; Ho, J.S.; Chen, L.Y.; Poon, A.S. Wireless Power Transfer to a Cardiac Implant. *Appl. Phys. Lett.* **2012**, *101*, 073701. [CrossRef]
10. Xu, Q.; Hu, D.; Duan, B.; He, J. A Fully Implantable Stimulator with Wireless Power and Data Transmission for Experimental Investigation of Epidural Spinal Cord Stimulation. *IEEE Trans. Neural Syst. Rehabil. Eng.* **2015**, *23*, 683–692. [CrossRef] [PubMed]
11. Karuppuswamy, R.; Arumugam, K. Folded Architecture for Digital Gammatone Filter Used in Speech Processor of Cochlear Implant. *ETRI J.* **2013**, *35*, 697–705. [CrossRef]
12. Zhao, Y.; Nandra, M.; Yu, C.C.; Tai, Y.C. High Performance 3-coil wireless Power Transfer System for the 512-electrode Epiretinal Prosthesis. In Proceedings of the 2012 IEEE Annual International Conference Engineering in Medicine and Biology Society (EMBC), San Diego, CA, USA, 28 August–1 September 2012.
13. Gutruf, P.; Rogers, J.A. Implantable, Wireless Device Platforms for Neuroscience Research. *Curr. Opin. Neurobiol.* **2018**, *50*, 42–49. [CrossRef] [PubMed]
14. Oxford Sandy and Black Pigs Classed as a Rare Breed Pig and Are Now Recognised with the RBST. Available online: http://www.oxfordsandyblackpigs.org.uk (accessed on 7 January 2019).
15. Webster, J.G. *Medical Instrumentation: Application and Design*, 4th ed.; Wiley: Hoboken, NJ, USA, 2009.
16. Medical Device Radiocommunications Service (MedRadio). Available online: https://www.fcc.gov/medical-device-radiocommunications-service-medradio (accessed on 7 January 2019).
17. Schuylenbergh, K.V.; Puers, R. An Introduction on Telemetry. In *Inductive Powering*; Springer: Berlin/Heidelberg, Germany, 2009; pp. 21–28.
18. Seo, D.-W.; Lee, J.-H.; Lee, H. Integration of Resonant Coil for Wireless Power Transfer and Implantable Antenna for Signal Transfer. *Int. J. Antennas Propag.* **2016**, *2016*, 7101207. [CrossRef]
19. Lee, J.-H.; Seo, D.-W. Design of Miniaturized and Bandwidth-Enhanced Implantable Antenna on Dielectric/Ferrite Substrate for Wireless Biotelemetry. *IEICE Trans. Commun.* **2017**, *100*, 227–233. [CrossRef]
20. Institute of Electrical and Electronics Engineers (IEEE). *IEEE Standard for Safety Levels with Respect to Human Exposure to Radiofrequency Electromagnetic Fields, 3 kHz to 300 GHz*; IEEE Std C95.1; IEEE: Piscataway Township, NJ, USA, 1999.
21. Scholten, K.; Meng, E. Materials for Microfabricated Implantable Devices: A Review. *Lab Chip* **2015**, *15*, 4256–4272. [CrossRef] [PubMed]
22. Celik, N.; Manivannan, N.; Strucdwick, A.; Balachandran, W. Graphaene-Enabled Electrodes for Electrocardiogram Monitoring. *Nanomaterials* **2016**, *6*, 156. [CrossRef] [PubMed]

Article

Tongue Pressure Sensing Array Integrated with a System-on-Chip Embedded in a Mandibular Advancement Splint

Yun-Ting Chen [1], Kun-Ying Yeh [2], Szu-Han Chen [3], Chuang-Yin Wang [1], Chao-Chi Yeh [1], Ming-Xin Xu [1], Shey-Shi Lu [2], Yunn-Jy Chen [3] and Yao-Joe Yang [1,*]

[1] Department of Mechanical Engineering, National Taiwan University, Taipei 10617, Taiwan; ytchen@mems.me.ntu.edu.tw (Y.-T.C.); wang2333@mems.me.ntu.edu.tw (C.-Y.W.); scott@mems.me.ntu.edu.tw (C.-C.Y.); xmx@mems.me.ntu.edu.tw (M.-X.X.)

[2] Department of Electronics Engineering, National Taiwan University, Taipei 10617, Taiwan; d98943028@ntu.edu.tw (K.-Y.Y.); sslu@ntu.edu.tw (S.-S.L.)

[3] Department of Dentistry, National Taiwan University Hospital, Taipei 10048, Taiwan; schweitzerchen@hotmail.com (S.-H.C.); chenyj@ntu.edu.tw (Y.-J.C.)

* Correspondence: yjy@ntu.edu.tw; Tel.: +886-2-3366-2712

Received: 6 June 2018; Accepted: 10 July 2018; Published: 14 July 2018

Abstract: Obstructive sleep apnea (OSA), which is caused by obstructions of the upper airway, is a syndrome with rising prevalence. Mandibular advancement splints (MAS) are oral appliances for potential treatment of OSA. This work proposes a highly-sensitive pressure sensing array integrated with a system-on-chip (SoC) embedded in a MAS. The device aims to measure tongue pressure distribution in order to determine the efficacy of the MAS for treating OSA. The flexible sensing array consists of an interdigital electrode pair array assembled with conductive polymer films and an SoC capable of retrieving/storing data during sleep, and transmitting data for analysis after sleep monitoring. The surfaces of the conductive polymer films were patterned with microdomed structures, which effectively increased the sensitivity and reduced the pressure sensing response time. The measured results also show that the crosstalk effect between the sensing elements of the array was negligible. The sensitivity of the sensing array changed minimally after the device was submerged in water for up to 100 h.

Keywords: obstructive sleep apnea; mandibular advancement splint; pressure sensing array; conductive polymer; system-on-chip

---

1. Introduction

Epidemiological studies published during the past decade have shown that the prevalence of obstructive sleep apnea (OSA) was, on average, about 22% in men and 17% in women. Some studies have also shown that the prevalence has increased significantly (from 2008 to 2013, a 37% increase for men and a 50% increase for women) [1]. Hypercapnia, hypoxemia, and sleep fragmentation are typical OSA symptoms and are caused by the collapse of the upper airway while sleeping [2]. Patients with OSA are at higher risk of being involved in a vehicle crash due to OSA-related characteristics such as daytime sleepiness, high apnea-hypopnea indices (AHI), and low oxygen saturation levels [3]. Moreover, OSA may increase the possibility of fatal cardiovascular events [4].

Nasal continuous positive airway pressure (CPAP) is the most common treatment for OSA. Although health-related improvements with CPAP have been demonstrated, the inconvenience and discomfort of using CPAP has resulted in relatively low acceptance rate by patients, which limits its clinical effectiveness. In recent years, oral appliances, particularly mandibular advancement splints (MAS), which increase the airway lumen volume by shifting the positions of the jaw and tongue, have been developed as a potential alternative treatment for OSA [5]. While MAS have been demonstrated to be effective for some patients, they may be ineffective for other patients or even result in worse OSA syndromes. Therefore, various research teams around the world have reported studies on the fundamental mechanism and efficacy of MAS therapy for OSA.

In general, the change in the airway lumen volume caused by using MAS can be observed by using tools such as endoscopy [6,7], lateral cephalometric radiography [8,9], fluoroscopy [10,11], computed tomography (CT) [12,13], and magnetic resonance imaging (MRI) [14,15]. In Ref. [6], Ryan et al. studied the correlation between the improvement in OSA symptoms and the changes in pharyngeal dimensions produced by a MAS by videoendoscopy of the upper airway. Mehta et al. [8] conducted a randomized controlled study of MAS for OSA using lateral cephalometric radiography. This study suggests that treatment outcome of MAS or OSA can be predicted by a combination of anthropomorphic, polysomnographic, and radiographic measurements. In Ref. [10], lateral cephalometric radiography and fluoroscopy were demonstrated to be effective at evaluating if the advancement of the mandibular will increase the oropharyngeal airway of a patient. In addition, cine CT was used to investigate the effect of MAS on pharyngeal size and shape [12]. Cross-sectional CT images recorded pharyngeal changes for two cycles of respiration. Sanner et al. [14] presented a method for predicting the efficacy of MAS for OSA by using ultrafast MRI to record the pharynx images for patients during both transnasal shallow respiration and performance of the Muller maneuver. These tools and methods are well-established and capable of acquiring accurate volumetric pharyngeal images. However, some of them are quite expensive or bulky; therefore, the measurements can only be performed in hospitals. As a result, these tools are not suitable for conducting measurement of OSA patients during sleep.

This work proposes embedding a tactile sensing array in a MAS to monitor the position and force of a patient's tongue during sleep. The sensing device, which was fabricated by micromachining techniques, consisted of a patterned conductive polydimethylsiloxane (PDMS) film and a flexible printed circuit board (FPCB) with an array of interdigital electrode pairs. Multi-wall carbon nanotubes (MWCNT) were employed as the conductive fillers in the PDMS film. Also, microdomed structures were transferred onto the surface of the PDMS film using a nylon membrane filter as the mold. The sensor array was connected to a system-on-chip (SoC) capable of retrieving/storing data during sleep, and transmitting data for analysis after sleep monitoring.

2. Device Design and Operational Principles

2.1. Sensor Array Principle and Design

During sleep, gravity may contribute to the falling back of the soft palate and tongue; this in turn narrows the upper airway and results in OSA symptoms [16], as shown in Figure 1a,b. Therefore, detecting the tongue force imposed on the palate as well as the position of the tongue during sleep may help improve MAS when treating OSA [17]. In this work, the proposed tactile sensing array embedded in a MAS is capable of measuring these physical quantities. Figure 2a illustrates how the pressure force distribution exerted on the palate during sleep can be measured by the proposed sensing array on the MAS. Figure 2b shows that the falling back of the tongue due to gravity in the supine posture can be readily detected by the sensing array.

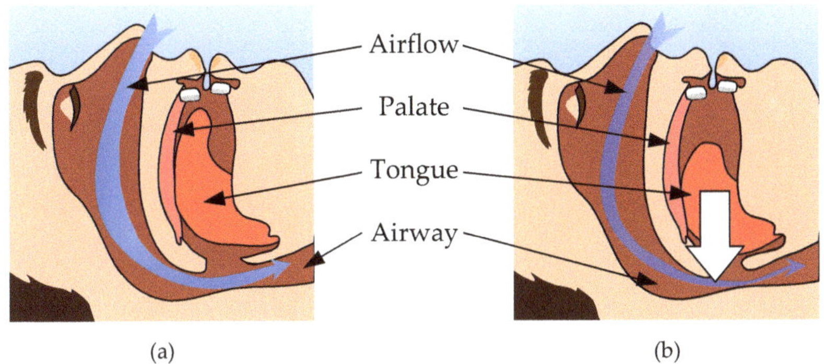

Figure 1. Illustration of the tongue falling back due to gravity. (**a**) Patient's tongue touches onto palate when sleeping. (**b**) Patient's tongue falls back due to gravity.

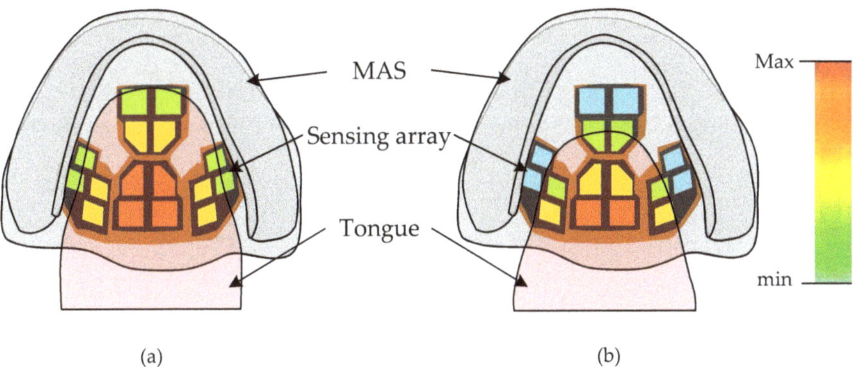

Figure 2. Working principle of device: (**a**) Tongue pressure distribution measured by the sensing array. (**b**) Change of pressure distribution after the tongue falls back.

Figure 3a,b show exploded and schematic views, respectively, of the proposed tongue pressure-sensing device. As shown in Figure 3a, the sensing array consisted of four PDMS-MWCNT polymer films and an FPCB with an array of interdigital electrode pairs (Flex PCB, Ping-Yi Electronics Corporation, Taoyuan, Taiwan). By using a nylon membrane filter as the mold with numerous micropores, microdomed structures were patterned on the surface of the polymer films. These randomly distributed microdomed structures, which have an average base width of approximately 4.5 µm, are essential to achieve high sensitivity for the pressure sensing element. Figure 3b shows the assembled device embedded in a MAS.

Figure 3c shows a cross-sectional view of the contacting interface between the polymer film and the interdigital electrode pair of the FPCB (without the SoC circuit). As uniform pressure is applied, as shown in the right part of Figure 3c, the microdomed structures are deformed, which in turn causes a rapid increase in the contact surface area between the conductive polymer and the planar electrode. Therefore, the contact resistance decreases sharply as the applied pressure increases. This sharp resistance change induced by small external forces is the so-called tunneling piezoresistive effect [18]. Devices with this type of piezoresistive effect are much more sensitive than typical conductive polymer-based devices [19].

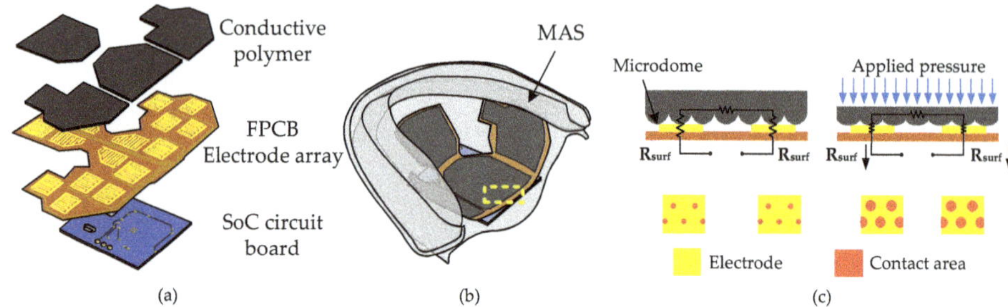

Figure 3. Diagrams of the tongue pressure-sensing device: (**a**) exploded view of the sensing array, (**b**) schematic of the device embedded in a mandibular advancement splint (MAS), and (**c**) cross-sectional view of the sensing element before and after pressure force is applied.

2.2. SoC Architecture

The SoC attached to the sensing array is capable of retrieving the detected signal from the sensing array, wirelessly recharging the Li-ion battery, and wirelessly transmitting signals. A block diagram of the SoC is shown in Figure 4. The system consists of an analog front-end (AFE) circuit, a 10-bit successive-approximation-register (SAR) analog-to-digital converter (ADC), a low-power transmitter, and a digital processor, which includes a digital micro-controller unit (MCU) and a power management unit (PMU).

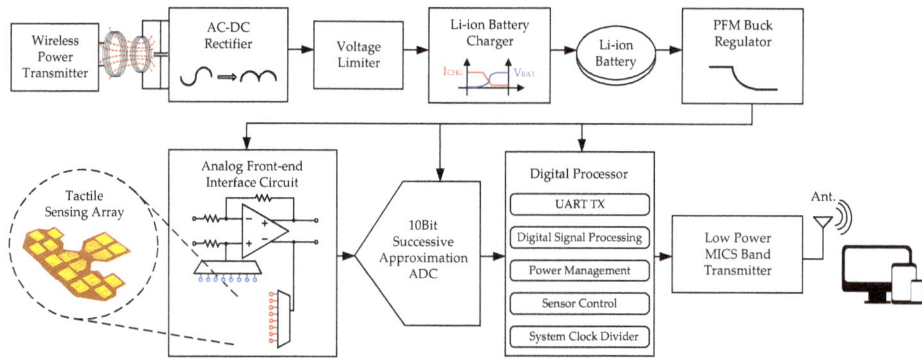

Figure 4. System-on-chip (SoC) block diagram.

The AFE circuit utilizes a low-power operational amplifier to form a resistor feedback amplifier with a loop gain defined by the ratio of resistors. The SAR ADC is utilized in the proposed system because of its advantages such as low power, high resolution, and low topological complexity. The SAR ADC is only suitable for operating at a relatively low frequency, which is appropriate in biomedical applications. The operation of the SAR ADC is based on a binary-search algorithm, which finds the final interval by comparing the sampled input signal with the voltages of $V_{ref}/2^n$ sequentially.

3. Fabrication

3.1. Fabrication of the Conductive Polymer

The conductive polymer was a composite of MWCNT and PDMS. The prepolymer for the PDMS-MWCNT composite was prepared by dispersing MWCNT (Golden Innovation Business Corp., New Taipei, Taiwan) with diameters of approximately 20 nm and lengths ranging from 10 to 50 μm into hexane and mixing with PDMS prepolymer (Sylgard 184A, Dow Corning Corp., Midland,

MI, USA) at a ratio of 4:1. The mixture was blended with a stirrer for 2 h. The concentration of MWCNT in the PDMS-MWCNT prepolymer was about 6 wt %. Note that the hexane served as the dispersant for improving prepolymer uniformity during blending. Then, the curing agent (Sylgard 184B, Dow Corning Corp., Midland, MI, USA) was dispersed with the PDMS-MWCNT prepolymer at a ratio of 10:1 and stirred for 50 min. After degassing in a chamber for 120 min, the PDMS-MWCNT prepolymer was ready for the subsequent soft-lithography process.

Figure 5 illustrates the fabrication process of the conductive polymer with microdomed structures. First, a layer of 200-µm thick-film SU-8 photoresist (SU-8 2050, MicroChem Co., Westborough, MA, USA) was spin-coated onto a silicon handling wafer, as shown in Figure 5a. Then, a nylon membrane filter (MS® nylon membrane filter, Membrane Solutions Corp., Kent, WA, USA) with a pore size of 3 µm was placed on the top surface of the SU-8 layer, as shown in Figure 5b. Next, a second layer of 200-µm SU-8 was spin-coated (Figure 5c) and patterned with a dose of 300 mJ/cm^2 (Figure 5d). As shown in Figure 5e, the SU-8 mold was formed after SU-8 development. Then, a layer of 200-µm PDMS-MWCNT prepolymer was added into the mold (Figure 5f) and cured at 95 °C for 5 h. After removing the SU-8 mold with a stripper (Remover PG, MicroChem Co., Westborough, MA, USA) (Figure 5g), the patterned conductive polymer was peeled from the membrane filter (Figure 5h), and the patterns on the nylon membrane filter were transferred to the conductive polymer. Figure 6a,b show scanning electron microscopy (SEM) images of the surfaces of the nylon membrane filter and the patterned conductive polymer, respectively. The conductive polymer films were then placed to the top of the electrode array on the FPCB. The edges of polymer films were sealed by using PDMS prepolymer.

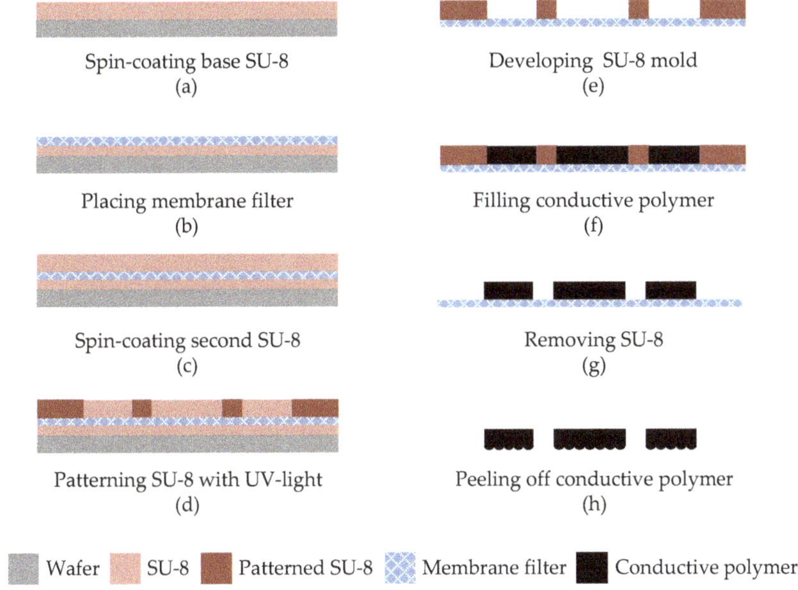

Figure 5. The fabrication process of the conductive polymer with microdomed structures.

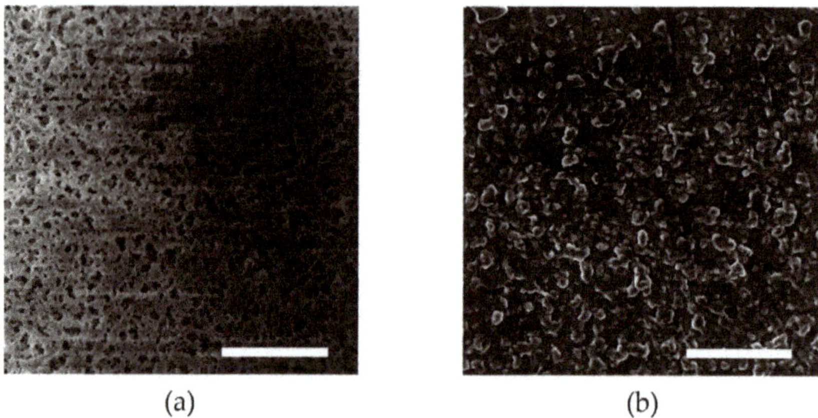

(a)   (b)

Figure 6. SEM pictures of the (**a**) nylon membrane filter and (**b**) conductive polymer. The scale bars indicate a length of 50 µm.

3.2. Assembling and Packaging

Figure 7 shows mandible and maxilla molds on which a MAS was attached. Figure 7a shows the MAS attached to the maxilla mold, and Figure 7b shows the MAS attached to the mandible mold. The proposed device (i.e., the sensing array with the SoC circuit board) was glued to the palatal plate of a MAS using denture resin (Hygenic® Denture Resin, Coltene Corp., Madrid, Spain). The flexible sensing array was assembled with the conductive polymer films with microdomed structures. The shape of the sensing array was designed to allow the sensing array to fit onto the curved roof of the MAS as closely as possible. The sensing array was composed of 16 electrode pairs. Each electrode finger was 400 µm wide, and the gap between fingers was 200 µm. The total sensing area of the interdigital electrode array was 477 mm^2. The SoC circuit board was connected to the sensing array by soldering a 16-wire flexible flat cable between them. The PDMS prepolymer was applied to the solder spot for both strength and isolation. The entire system was coated with a 3-µm biocompatible Parylene-C film (Parylene-C, La Chi Enterprise Corp., Taipei, Taiwan).

(a)   (b)

Figure 7. The MAS attached to the (**a**) maxilla mold and (**b**) mandible mold.

Figure 8a shows a posterior view of the maxilla and mandible molds with the proposed device attached. This picture also shows that the flexibility of the proposed sensing array device facilitates its installation onto a customized MAS. Figure 8b shows a side view of the molds with the proposed device attached.

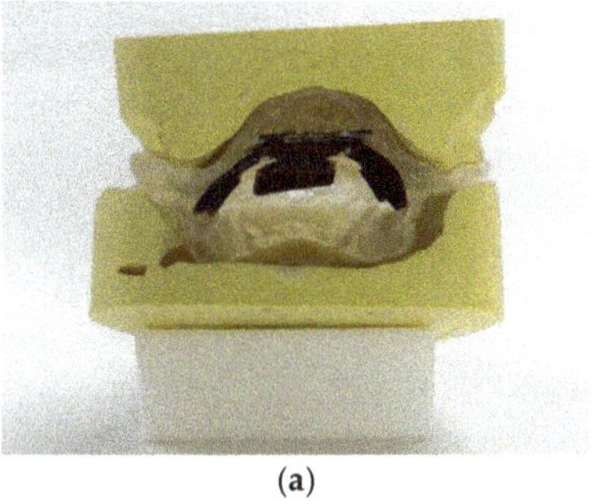 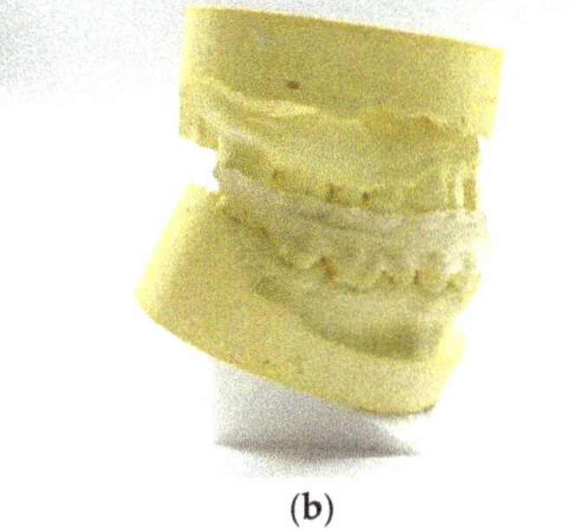

(**a**) (**b**)

Figure 8. (**a**) Posterior view of the maxilla and mandible molds with the proposed device attached. (**b**) Side view of the molds with the proposed device attached.

4. Measurements

The experimental setup for measuring the piezoresistive characteristics and dynamic responses of the sensing elements are shown in Figure 9a,b, respectively. As shown in Figure 9a, the force applied to the sensing element was measured using a force gauge (HF-1, resolution: 1 mN, ALGOL Instrument Corp., Taoyuan, Taiwan) attached to a vertical translation stage (SR-300C, displacement resolution: 1 μm, Dispenser Tech Corp., New Taipei, Taiwan). A circular polymethylmethacrylate (PMMA) rod of 7.4 mm in diameter was glued on the sensing head of the force gauge. The resistance of the sensing element was measured using a source meter (Model 2400, Keithley Instruments Corp., Solon, OH, USA). Figure 9b shows the experimental setup for measuring the dynamic responses of the sensing element. A piezoelectric actuator (P-621.20L, Physik Instruments Corp., Karlsruhe, Germany) was driven by a power amplifier fed with square wave signals generated by a function generator (GFG8216A, GW Instek, New Taipei, Taiwan). The sensing element was periodically pressed and released by a PMMA block attached to the head of the actuator. The contact area of the PMMA block is 3 mm × 3 mm. A laser Doppler interferometer was employed to measure the displacement of the actuator. The transient behaviors of the sensing element were retrieved using a simple circuit which converted the resistance of the sensing element into a voltage output.

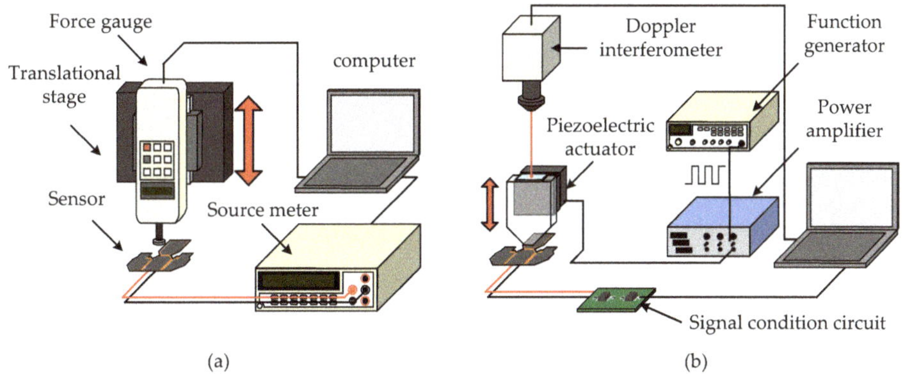

Figure 9. (**a**) Experimental setup for measuring the piezoresistive characteristics of the sensing element. (**b**) Experimental setup for measuring the dynamic responses of the sensing element.

Sensing elements with electrode finger widths of 300 μm (Device A) and 400 μm (Device B) were fabricated and measured. The gap between fingers was 200 μm for both devices. Figure 10 shows the results of the pressure-resistance measurement of the sensing elements. The results indicate that the proposed devices exhibited high sensitivities at lower pressure region (below 10 kPa). It is worth noting that the sensitivity of Device B was larger than that of Device A because Device B had a larger effective electrode area due to its larger finger width. Table 1 shows the sensitivities calculated for devices with various finger widths. Figure 11 shows the measured hysteresis effects of the fabricated sensing element. The external pressure force slowly increased from zero to 70 kPa, and then retreated to zero. As shown in the figure, the hysteresis effect was not significant. It is possibly because the piezoresistive behavior of the proposed sensor is primarily caused by the sharp change in contact area between microdomed structures and the electrodes, which predominates the viscoelastic response observed in typical polymer-based piezoresistive material.

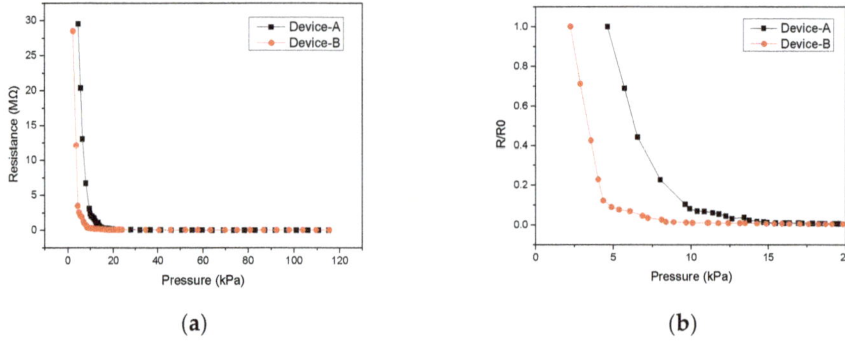

(a)   (b)

Figure 10. (a) Measured responses of the devices of various widths of electrode fingers. (b) Normalized measured responses at lower pressure region (0–20 kPa).

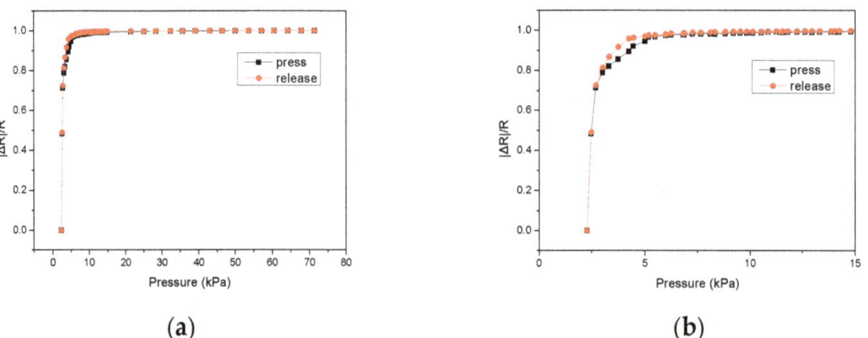

(a)   (b)

Figure 11. (a) Measured hysteresis curve of the sensing element. (b) The subset of (a) in lower pressure region.

Table 1. Sensitivities of the devices with different finger widths.

Devices	Device A	Device B
Finger width (μm)	300	400
Gap width (μm)	200	200
Sensitivity (kPa^{-1})	0.177	0.288

Figure 12 shows the measured dynamic responses of the sensing element as it was periodically pressed and released by the piezoelectric actuator. The sensor response lagged behind the motion of the piezoelectric actuator by about 5 ms. Figure 13 depicts the results of repeatability and drift tests by measuring the sensing element for 2500 cycles using the same experimental setup as shown in Figure 9b driven with a 1-Hz square wave. Figure 13b is the subset snapshot of Figure 13a. The sensing element exhibited reasonably good repeatability and was capable of long-term pressure sensing.

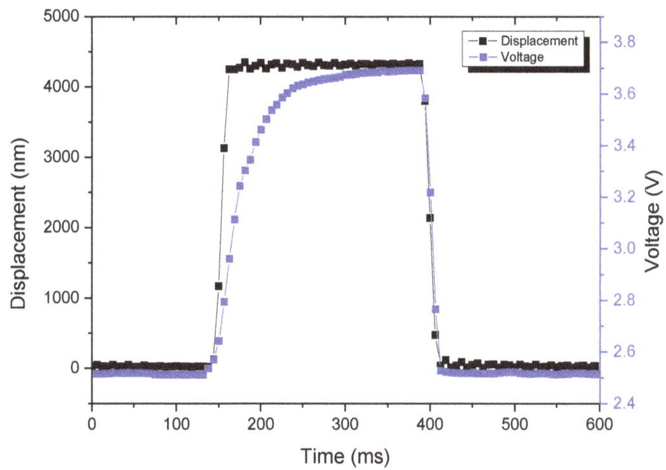

Figure 12. Dynamic square waves response of the sensing element.

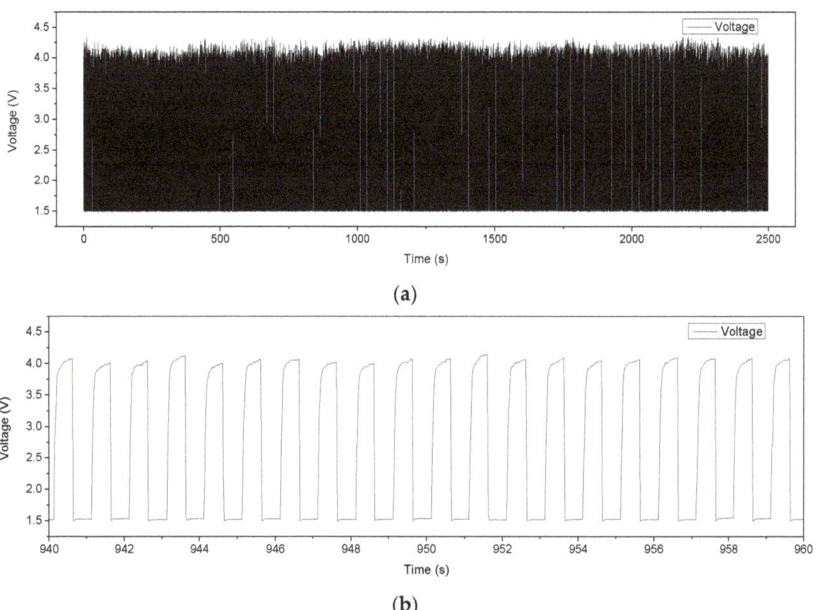

(a)

(b)

Figure 13. Repeatability test of the sensing element: (**a**) 2500 s of the sensing element's dynamic responses. (**b**) A snapshot of (**a**) between 940 s and 960 s.

Figure 14 shows the crosstalk effect measured between the sensing elements. An external pressure force was exerted on point X, and the resistances at points M, G, L, H, D, and F were measured separately. The measured results show that the resistance at point X decreased considerably as pressure force was applied to point X, while the resistances of the adjacent points (i.e., M, G, L, H, D, and F) were almost unchanged even when the applied force was raised to 125 kPa, which indicates that the crosstalk effect was not discernible for the proposed sensing array.

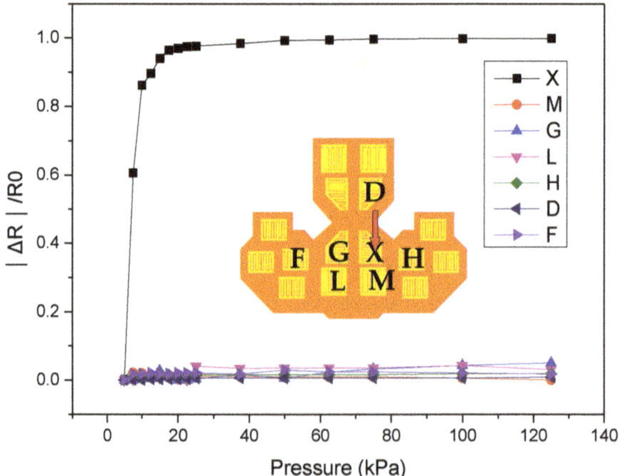

Figure 14. The measured normalized resistances for crosstalk effect measurements. The inset shows the position of the corresponding sensing element on the array.

Changes in the proposed device's sensitivity after the sensing array, which was packaged using Parylene-C, was submerged in deionized water for 50 h, 75 h, and 100 h were evaluated to demonstrate its water-resistant capability. As shown in Figure 15, the pressure-resistance curves of the sensing array after being submerged in water for various periods of time were almost the same as the original one, which indicates that the changes in sensitivity were negligible. This preliminary finding shows that Parylene-C coating can provide reasonably effective short-term water-resistant capability.

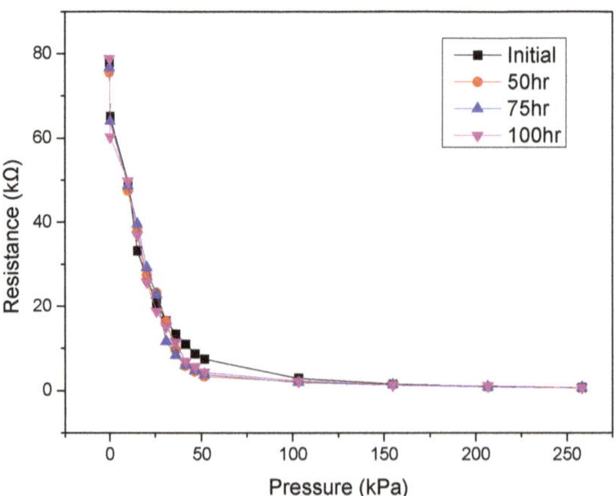

Figure 15. Pressure–resistance curves after the packaged device was submerged in water for different periods of time.

Figure 16 shows screenshots of the sensing array's responses to an artificial silicon tongue touching the sensing array installed on a MAS that was attached between mandible and maxilla molds. The insets in these sub-figures are the pressure distributions measured by the sensing array. The color-bar legend is from 5 kPa (cyan) to 50 kPa (red). Figure 16a,b show the results of when the artificial tongue pressed the center and posterior areas of the sensing array. Figure 16c,d show the results of the tongue pressing the sinistral and dextral areas of the sensing array.

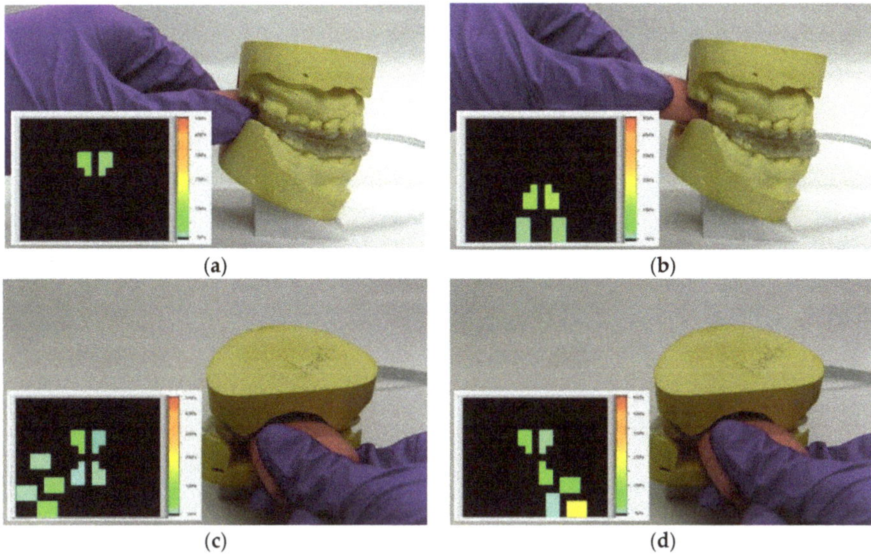

Figure 16. Screenshots of the sensing array measuring the pressure distribution of a silicon tongue. The tongue is pressed against the center (**a**), posterior (**b**), left (**c**), and right (**d**) areas of the sensing array.

To achieve a sufficiently compact module fitting in the very limited space of a MAS, a specifically designed SoC chip was realized by UMC 0.18 μm standard CMOS process (United Microelectronics Corp., Hsinchu, Taiwan). Figure 17 shows the measured input-output characteristic of AFE by feeding sinusoidal input signals with amplitudes ranging from 50 to 1800 mV. The blue line is the regression line of small input voltage with a calculated gain of 1.21. The output amplitude starts to be compressed when the input amplitude reaches around 850 mV. Figure 18 shows the measured ADC output codes corresponding to different DC input levels from the AFE. The DC input levels from the AFE were 900, 1000, 1100, 1200, 1300, 1400, 1500, 1600, 1620, and 1640 mV. The corresponding output codes of ADC in Figure 18 presented a quite linear characteristic. The glitches on these curves resulted from supply and environmental noises which can be simply eliminated using additional off-chip components. Figure 19 is the micrograph of the SoC implemented for the sensing array. The chip area, including the testing pads, is 3.16 mm × 3.09 mm. Table 2 lists the measured performance summary which concludes the detailed measurement results of all the sub-blocks in this SoC chip.

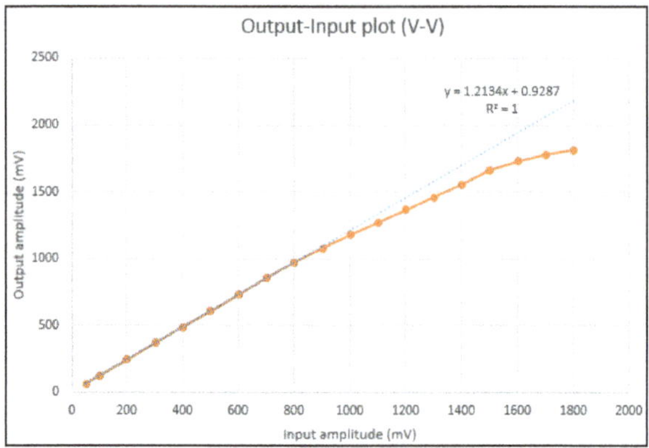

Figure 17. Measured input–output characteristic of the analog front-end (AFE).

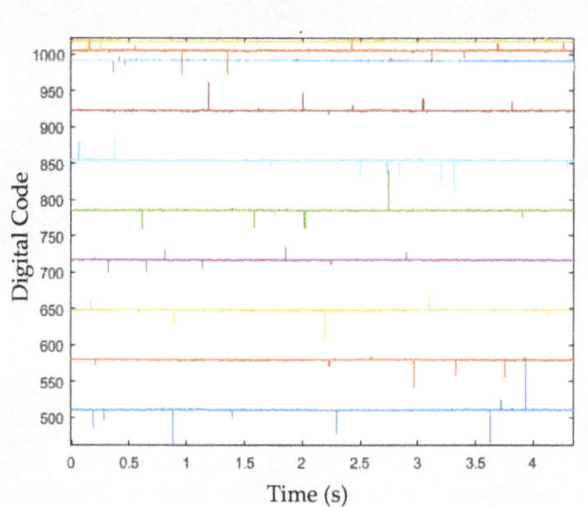

Figure 18. Measured analog-to-digital converter (ADC) output codes corresponding to different DC input levels from the AFE.

Figure 19. The micrograph of the System-on-Chip for the sensing array.

Table 2. Summary of measured performance of the SoC.

Summary	
Technology	UMC 0.18um CMOS
Chip Area	3.16×3.09 mm^2
Supply Voltage	1.8 V
Analog Front-end	
SNR	63.1 dB
GainBWProduct	4.4 MHz
Linearity	~0.999
Sensitivity	14 mV/kPa
Power Consumption	6 mW
SAR ADC	
Resolution	10 bit
Sampling Rate	200 kS/s
SFDR	74.0 dB@99.8 kHz
DNL	+0.127/−0.18 LSB
INL	+0.30/−0.14 LSB
ENOB	9.47 bit@99.8 kHz
Power Consumption	6.24 mW
Digital Control Circuits	
Clock Rate	UART: 38.4 KHz; Scan rate: 1 Hz
Power Consumption	2.53 mW
Buck DC-DC Converter	
Input voltage	3.6 V
Power Consumption	6 mW @ 98 mA, η = 78.3%
OOK Transmitter	
Carrier Frequency	403 MHz (MICS)
Output Power	−14 dBm
Power Consumption	1 mW
Off-chip Components	
Li-ion Battery	3.6 V/50 mAh
Power Consumption	EEPROM: 0.72 mV @ Read; 9 mV @ Write

5. Conclusions

This work presented a highly-sensitive pressure sensing array integrated with an SoC circuit embedded in a MAS for measuring tongue pressure distribution. The proposed flexible device consists of an FPCB with an interdigital electrode array with conductive polymer films and an SoC for retrieving, storing, and transmitting data. The pressure sensing array exhibited high sensitivity and rapid response because of the microdomed structures patterned onto the conductive polymer films. The crosstalk effect between adjacent sensing elements was not obvious. In addition, the device displayed effective short-term water-resistant capability. The proposed device can be used to assist physicians to study and improve the efficacy of MAS for OSA.

Author Contributions: The sensing array was designed and fabricated by Y.-T.C., C.-C.Y. and M.-X.X. in Y.-J.Y.'s research group. The SoC was designed by K.-Y.Y. and S.-S.L. The measurement setup was designed and implemented by Y.-T.C. and C.-Y.W. The paper was written by Y.-T.C. and Y.-J.Y. The medical oversights are provided by S.-H.C. and Y.-J.C. The principle investigator is Y.-J.Y.

Funding: This research was partially funded by the Ministry of Science and Technology, Taiwan grant number 105-2221-E-002-126.

Acknowledgments: The authors are grateful for Fu Tan's help on the measurement using an interferometer.

Conflicts of Interest: The authors declare no conflicts of interest.

References

1. Franklin, K.A.; Lindberg, E. Obstructive sleep apnea is a common disorder in the population—A review on the epidemiology of sleep apnea. *J. Thorac. Dis.* **2015**, *7*, 1311. [PubMed]

2. Epstein, L.J.; Kristo, D.; Strollo, P.J.; Friedman, N.; Malhotra, A.; Patil, S.P.; Ramar, K.; Rogers, R.; Schwab, R.J.; Weaver, E.M.; et al. Clinical guideline for the evaluation, management and long-term care of obstructive sleep apnea in adults. *J. Clin. Sleep Med.* **2009**, *5*, 263. [PubMed]

3. Tregear, S.; Reston, J.; Schoelles, K.; Phillips, B. Obstructive sleep apnea and risk of motor vehicle crash: Systematic review and meta-analysis. *J. Clin. Sleep Med.* **2009**, *5*, 573. [PubMed]

4. Marin, J.M.; Carrizo, S.J.; Vicente, E.; Agusti, A.G. Long-term cardiovascular outcomes in men with obstructive sleep apnoea-hypopnoea with or without treatment with continuous positive airway pressure: An observational study. *Lancet* **2005**, *365*, 1046–1053. [CrossRef]

5. Schmidt-Nowara, W.; Lowe, A.; Wiegand, L.; Cartwright, R.; Perez-Guerra, F.; Menn, S. Oral appliances for the treatment of snoring and obstructive sleep apnea: A review. *Sleep* **1995**, *18*, 501–510. [CrossRef] [PubMed]

6. Ryan, C.F.; Love, L.L.; Peat, D.; Fleetham, J.A.; Lowe, A.A. Mandibular advancement oral appliance therapy for obstructive sleep apnoea: Effect on awake calibre of the velopharynx. *Thorax* **1999**, *54*, 972–977. [CrossRef] [PubMed]

7. Schessl, J.; Rose, E.; Korinthenberg, R.; Henschen, M. Severe obstructive sleep apnea alleviated by oral appliance in a three-year-old Boy. *Respiration* **2008**, *76*, 112–116. [CrossRef] [PubMed]

8. Mehta, A.; Qian, J.; Petocz, P.; Darendeliler, M.A.; Cistulli, P.A. A Randomized, controlled study of a mandibular advancement splint for obstructive sleep apnea. *Am. J. Respir. Crit. Care Med.* **2001**, *163*, 1457–1461. [CrossRef] [PubMed]

9. Robertson, C.; Herbison, P.; Harkness, M. Dental and occlusal changes during mandibular advancement splint therapy in sleep disordered patients. *Eur. J. Orthod.* **2003**, *25*, 371–376. [CrossRef] [PubMed]

10. Battagel, J.M.; L'Estrange, P.R.; Nolan, P.; Harkness, B. The role of lateral cephalometric radiography and fluoroscopy in assessing mandibular advancement in sleep-related disorders. *Eur. J. Orthod.* **1998**, *20*, 121–132. [CrossRef] [PubMed]

11. Lee, C.H.; Kim, J.W.; Lee, H.J.; Seo, B.S.; Yun, P.Y.; Kim, D.Y.; Yoon, I.Y.; Rhee, C.S.; Park, J.W.; Mo, J.H. Determinants of treatment outcome after use of the mandibular advancement device in patients with obstructive sleep apnea. *Arch. Otolaryngol. Head. Neck. Surg.* **2010**, *136*, 677–681. [CrossRef] [PubMed]

12. Kyung, S.H.; Park, Y.C.; Pae, E.K. Obstructive sleep apnea patients with the oral appliance experience pharyngeal size and shape changes in three dimensions. *Angle Orthod.* **2005**, *75*, 15–22. [PubMed]

13. Haskell, J.A.; McCrillis, J.; Haskell, B.S.; Scheetz, J.P.; Scarfe, W.C.; Farman, A.G. Effects of Mandibular Advancement Device (MAD) on Airway Dimensions Assessed with Cone-beam Computed Tomography. In *Seminars in Orthodontics*; Elsevier: New York, NY, USA, 2009; Volume 15, pp. 132–158.

14. Sanner, B.M.; Heise, M.; Knoben, B.; Machnick, M.; Laufer, U.; Kikuth, R.; Zidek, W.; Hellmich, B. MRI of the pharynx and treatment efficacy of a mandibular advancement device in obstructive sleep apnoea syndrome. *Eur. J. Orthod.* **2002**, *20*, 143–150. [CrossRef]

15. Chan, A.S.; Sutherland, K.; Schwab, R.J.; Zeng, B.; Petocz, P.; Lee, R.W.; Darendeliler, M.A.; Cistulli, P.A. The effect of mandibular advancement on upper airway structure in obstructive sleep apnoea. *Thorax* **2010**, *65*, 726–732. [CrossRef] [PubMed]

16. Yildirim, N.; Fitzpatrick, M.F.; Whyte, K.F.; Jalleh, R.; Wightman, A.J.; Douglas, N.J. The Effect of posture on upper airway dimensions in normal subjects and in patients with the sleep apnea/hypopnea syndrome. *Am. Rev. Respir. Dis.* **1991**, *144*, 845–847. [CrossRef] [PubMed]

17. Chen, Y.T.; Yeh, C.C.; Chen, S.H.; Xu, M.X.; Tan, F.; Chen, Y.J.; Yang, Y.J. A Mandibular Advancement Device Embedded with Polymer-based Tunneling Piezoresistive Pressure Sensing Array. In Proceedings of the 31st IEEE International Conference on Micro Electro Mechanical Systems (MEMS), Belfast, UK, 21–25 January 2018; pp. 51–54.

18. Park, J.; Lee, Y.; Hong, J.; Ha, M.; Jung, Y.-D.; Lim, H.; Kim, S.Y.; Ko, H. Giant tunneling piezoresistance of composite elastomers with interlocked microdome arrays for ultrasensitive and multimodal electronic skins. *ACS Nano* **2014**, *8*, 4689–4697. [CrossRef] [PubMed]
19. Ma, C.W.; Chang, C.M.; Lin, T.H.; Yang, Y.J. Highly sensitive tactile sensing array realized using a novel fabrication process with membrane filters. *J. Microelectromech. Syst.* **2015**, *24*, 2062–2070. [CrossRef]

MDPI

St. Alban-Anlage 66

4052 Basel

Switzerland

Tel. +41 61 683 77 34

Fax +41 61 302 89 18

www.mdpi.com

Micromachines Editorial Office

E-mail: micromachines@mdpi.com

www.mdpi.com/journal/micromachines